A Vision of Future Space Transportation

A Visual Guide to Future Spacecraft Concepts

All rights reserved under article two of the Berne Copyright Convention (1971).
We acknowledge the financial support of the Government to Canada through the
Book Publishing Industry Development Program for our publishing activities.
Published by Apogee Books an imprint of Collector's Guide Publishing Inc., Box 62034, Burlington, Ontario, Canada, L7R 4K2
Printed and bound in Canada
A Vision of Space Transportation: A Visual Guide to the Spacecraft of Tomorrow
by Tim McElyea
(Contributing Editor - Robert Godwin)
ISBN 1-896522-93-9
ISSN 1496-6921
Apogee Books Space Series
©2003 Apogee Books

A Vision of Future Space Transportation

A Visual Guide to Future Spacecraft Concepts

by

Tim McElyea

(Contributing Editor – Robert Godwin)

Author's Note

I see myself as an artist both by trade and passion. I am not a scientist and I don't pretend to be one. It just happens to be that growing up in the Rocket City, I have been around space exploration all of my life. I had space toys as a child and went to the Alabama Space and Rocket Center on elementary school field trips. My parents took the family to Florida to watch the first Apollo launches. The space program was a big part of my childhood. My fascination with space exploration started at an early age and continues today. Creating images for NASA is a dream come true for me and something that I am very excited to share.

This book does not pretend to depict every concept under consideration today. The concepts included here are some that I have had the opportunity to work with over the past years at some level. The concepts in this book are deliberately arranged in no particular order. It is not my intent to promote one project over another. Some concepts in this book have numerous images while others may have only one. The amount of coverage here is not a reflection on the viability of the concept. Some concepts are quite possible, while others are not seriously considered any longer. Every concept in this book has technical merit and some scientific foundation and that is why it is included. With each technology there are those who believe in it and are dedicating their professional career to it. On the other side are scientists who have reservations about its feasibility and may even question the basic physics behind it. By including some concepts and not including others I am not trying to promote one concept over another. This book is a sampling of just some of what NASA and industry are considering and it is not to be considered an endorsement.

In writing this book I set out to present an overview of what is envisioned for space transportation. I don't get lost in details and I try to stay above the controversy and disputes that are inevitable when a technological revolution is brewing. I rely heavily on the artwork, as I believe it conveys the passion and energy of space exploration. I let artistic license override any controversy and skepticism. It is my hope that this book will help spawn a new or renewed interest in space for someone.

Tim McElyea

Interactive CD References

Throughout the book, the following icons may appear at a vehicle topic heading. When you see these icons, refer to the included CD to learn more about the vehicle.

Refer to the CD to view the vehicle in 3D

Refer to the CD to see a digital video of the vehicle

Acknowledgements

In writing this book I relied upon discussions and interviews with various scientists and program leaders. I would not have attempted to write this book without their encouragement and willingness to take time to talk with me about their projects.

A special thanks goes to Steve Cook, Les Johnson, John Cole, David Harris, Michelle Munk, Dennis Smith, Rose Allen, Randy Baggett, Greg Garbe, Jonathan Jones, Melody Herman, Rae Ann Meyer, Michael Houts, Ron Litchford, Susan Turner, John London, and many others.

I wish to express my gratitude to everyone who supported this effort and our ongoing activities, facilitating meetings, and offering advice, content, and direction. Thanks to Joe Sparks, Debbie Scrivner, Mike Crabb, and Bruce Shelton.

Thanks to all of the helpful and inspirational people at Marshall Space Flight Center in the NGLT and OSP offices. Thanks for the energy, support, and the seemingly endless knowledge and resources of the people at SAIC and ASI.

Credits

Portions of this book are the products of a collaborative effort from a few individuals who took the time to make sure the information was correct. Any time a complex subject is simplified there is the strong likelihood of error and distortion. The significant efforts and writing contributions of Kirk Sorensen, Dr. Joseph Bonometti, Dr. Frank Curran, William Escher, and Dr. Thomas Roberts have helped to minimize inaccuracies and distortion of fact.

Proofing, commentary, and editing: Nicole Stephens, Darlene Smith, Holly Snow, Betty Roberts, Michelle Dietrich, and Rob Godwin.

The graphic designers and artists at Media Fusion, Inc. in Huntsville, Alabama created all of the illustrations in this book and CD. Illustrations were created by Tim McElyea, Kareem Marquez, Terri Reidy, Leroy Allen, and John Travis. The book design and layout is the work of Darlene Smith and Tim McElyea.

The multimedia CD was designed and created by Nicole Stephens, and Tim McElyea.

Content of the CD is the product of many producers including: Media Fusion, Inc., Boutwell Studios, SAIC, ASI, CSC, and ASRI. Individual contributors include the Media Fusion team mentioned above plus Holly Snow, Rick Smith, Austin Boyd, and John Dumoulin.

The Space Day chapter text was contributed by Ralph K. Coppola, Ed.D., Selma Mead, Karen Offringa, and Merri Oxley.

Artwork by Adam and Alycia McElyea, age 5.

A Vision of Future Space Transportation

This book is dedicated to the future explorers.

Contents

Foreword

Ever since people first looked up and were awed by the freedom and speedy grace of winged creatures, we have dreamed of flying. Over the ages, various humans – many of them labelled "crackpots" – tried to join the birds. They created kites, hot air balloons, zeppelins and gliders to glide on currents of air. Then, on December 17, 1903, the first powered, heavier-than-air vehicle lifted a person and launched a revolution. Since the Wright Brothers' remarkable experiment, we've pushed our frontiers from the Earth's surface through clouds and stratosphere, breaking barriers of sound and vacuum, exploring worlds those two pioneering brothers could barely have imagined. Moreover, we've done something equally miraculous – given the power of flights to millions, who now cruise the sky with little more thought than they would give to hailing a taxicab.

If the past 100 years are rightly considered the Century of Flight, will the next 100 years be the Century of Space? That seemed a certainty during the era of moon landings and "2001 A Space Odyssey." Only now some question whether our next step will be upward at all. The seventy years from Kitty Hawk to Voyager showed ever-greater achievements in speed and reach. Not so, the thirty years since.

Maybe it's time to remind ourselves. To rouse that sense of boldness and expectancy. To remember, once again, that we are a species of explorers.

As part of our re-awakening, this multimedia project will take you on a journey of discovery, through a galaxy of new transportation concepts designed to help us navigate through space and onto other worlds – concepts envisioned by some of the best minds in the aerospace industry.

At the same time, let's acknowledge that our next generation of aerospace professionals have much to contribute. This book also features some stellar creativity by elementary and middle school students who have tackled the challenges of living and working in space, designing their own aircraft and spacecraft of the future. It is to these budding inventors, aviators and explorers that we must turn as citizens of the Third Millennium begin to rise up, finding for humanity a place in the universe. Our destiny among the stars.

– David Brin

David Brin is a space scientist whose bestselling novels have won Hugo and Nebula awards and have been translated into 20 languages. His 1989 ecological thriller, *Earth*, foreshadowed global warming, cyberwarfare and the Web. A 1998 movie was loosely adapted from *The Postman*. His non-fiction book – *The Transparent Society* won the Freedom of Speech Award of the American Library Association.

*Not to know is bad;
    not to want to know is worse.*

— African Proverb

Chapter I

Why Space?

Chapter I
Why Space?

Why explore space? What's there for us? Why spend the energy and money in space when there are things we need to do here on Earth? These are valid questions with as many answers as there are questioners. It is human nature to question, challenge, and probe. While the purpose of this book is not to argue a case for space, I think it is important to at least touch on the topic from the start. After all, there is a significant effort being made to explore space, and the rewards and varied motives are not always commonly understood or shared.

When looking for a rationale for space exploration there are plenty of answers. Economic benefit is one of the first answers. Earth sciences and scientific research are some of the fundamental promises of the International Space Station. Pure scientific exploration is the obvious rationale for interplanetary missions. Military advantage is not always presented but with the ever-changing global political climate, we are seeing how space superiority provides a clear advantage for the war fighter. There is also space tourism. These are some of the mainstream motivations.

The Economic Benefits of Space

There are many businesses today realizing the benefits of space. Space provides an ideal relay point for communications. Satellites enable communications that would otherwise be impossible or impractical. The capabilities of communication satellites have created new businesses and expanded others. Likewise, satellite imagery has opened new markets. Then there is the business of providing access to space. These are some of the ways businesses are making money from space.

There are plenty of other economic incentives for space access but the current state of the industry is compounded by the global economic downturn that makes the case harder to argue. Presently, it is very expensive to do work in space. The section entitled "The Economic Barrier" describes how the current state of space access limits the economic potential. Before the potential of space can be unlocked the economic barrier must be broken. Paradoxically, it will

require a significant financial investment before the cost of access to space can be reduced.

Earth Sciences and Research

Space has literally changed the way we look at our planet. Space provides a perspective that cannot be found on the ground. Imaging satellites can watch the weather and help meteorologists predict weather patterns. Space imaging helps farmers and foresters understand their soil better and improve crop yields. Maps are more accurate because of space technology. Environmental scientists have an ever-increasing ability to evaluate the health of our home planet because of space. Space has real benefits that serve us every day and improves the quality of life here on Earth. There is a video clip on the CD that describes the benefits of space imaging further.

Many people think that there is no gravity in orbit. If that were true, satellites could not stay in orbit. The gravitational pull is very minimal compared to what it is on the ground. For this reason Low Earth Orbit (LEO) is said to have microgravity. This environment is unique to space. On Earth, gravity affects chemical and material manufacturing processes. Manufacturing processes that mix liquids, for example, have to fight gravity as gravity tries to pull heavier particles to the bottom of the mixture. In space, an almost perfect mix can be obtained. This means purer compounds and materials can be created. In the electronics and pharmaceutical industries, this can make a substantial difference in efficiency, performance, and effectiveness. Microgravity creates a laboratory environment that cannot be duplicated on the ground.

Scientific Exploration

The Space Age began in 1957 with the launch of Sputnik. It has been less than 50 years since then and we have only begun to understand our solar system. In the 1950's it was believed that Mars was covered with grass fields and canals. Today we know different. The Mars Global Surveyor is mapping the surface of Mars. Scientists believe that Mars has ice. With ice caps on

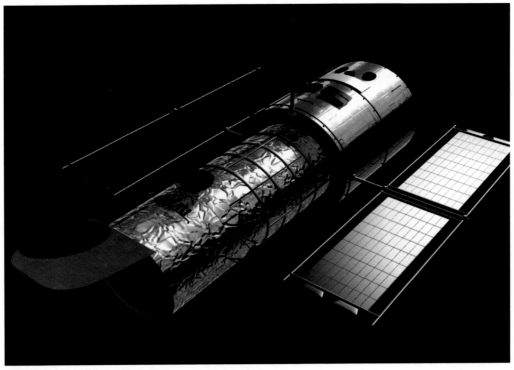

The Hubble space telescope.

Mars, the possibility of colonization is greatly increased. A colony on Mars brings with it much of the untold possibilities as did colonies of ancient times only on a much grander scale.

The Hubble space telescope has brought new discoveries that are more fantastic than science fiction. Space probes to the moons of Saturn or Jupiter may reveal the possibilities of life within our solar system. The moons Io and Titan have atmospheres and may have ice or water. If this proves true, they could contain life and be destinations for future missions. Today, NASA is more aggressively pursuing the search for life and evidence of life on other worlds.

Humans are explorers by nature. A toddler explores his environment at every opportunity. This is how he learns and grows. People climb mountains, explore the seas, and strive to learn and understand the world around us and beyond us because it is in our nature. It is

difficult to imagine our society where we no longer wish to learn more about our cosmic neighborhood.

Military in Space

NASA grew out of the Army's Ballistic Missile Command. The Army saw the military advantage of long-range rockets and has been investing in this technology since the middle of the last century. The global military climate is completely different than it was at the end of World War II. Today the military must have real-time intelligence in numerous theaters and be able to respond quickly. The U.S. military is retooling to be more flexible and more responsive in the face of new threats.

Information dominance on the battlefield is the new strategy for the military. Imaging satellites help locate, track, and target enemy forces and assets. Space hardware can detect launches and provide up-to-the-second information about enemy activities. A satellite in stationary orbit can keep a watchful eye on hot spots 24 hours a day. Current technology can provide imagery with resolution of objects smaller than 1 meter in detail. As imaging capabilities continue to increase, so will their value to the military strategists and planners.

Information is only useful when it can get into the hands of those who need it. The soldier in the field must have fast, secure, and reliable communications. Communication on the battlefield makes the difference between success and failure. When a military unit is rapidly deployed to a remote location, communication is not always readily available. In mountainous regions such as Afghanistan, the hilly terrain limits traditional radios and can leave the soldier isolated. New space communications can change this. The U.S. military is looking to develop a new satellite communications system that will allow the soldier cell-phone-like communications from any location in the world whether it is in the jungle, in a valley, or in the air. Currently, this level of communication does not exist and can only be done from space.

Hypersonic rockets hold the promise of rapid global strike. A hypersonic rocket could be launched from any location on Earth and hit a target anywhere on the planet within 90 minutes. When this capability is combined with the information dominance of advanced imaging and global communication, the military of the future will be able to discourage aggressors or eliminate them when required.

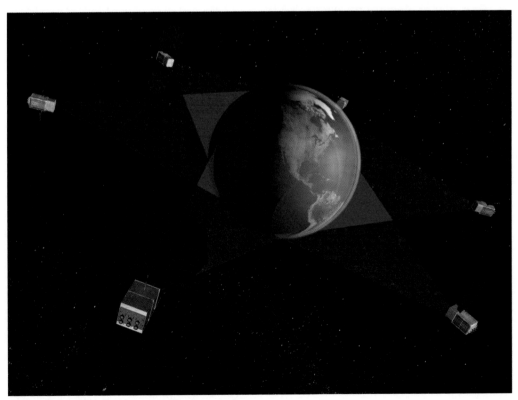

Military communication satellite network.

Space Tourism

I have always wanted to experience space flight. To be an astronaut was the first answer that I gave when asked what I wanted to be when I grew up. Today, the wealthy are willing to spend millions to experience space flight. Millionaires are lining up for a trip into space. Some private companies are currently booking flights on aircraft to take passengers up for a freefall flight to experience brief weightlessness. This is not at all the same thing as space flight but the desire to experience space is so strong that "close" is better than nothing.

If space travel can be made safer, more reliable and affordable, space tourism may one day become the largest industry for space. Just as the entertainment industry drives much of the computer industry, tourism may provide the revenue needed to create a market incentive and bring the cost of travelling to space down. Future spaceports may serve as departure points for vacation destinations such as an orbiting hotel or lunar trip. Needless to say, this is probably not going to be a reality in my lifetime but perhaps it will be for our children.

In the future, flights to space may be as routine as airflight is today.

Chapter II
Do We Really Need Something New?

*The best way
to escape from a problem
is to solve it.*

— Brendan Francis

Chapter II
Do We Really Need Something New?

Let's look at where we are today. We have the Space Shuttle and it is considered fairly routine. Today a shuttle launch hardly gets public attention and is barely newsworthy until something goes wrong. Steadily, we are building the International Space Station. We have sent probes to the outer solar system and robots to Mars. Satellite dishes are a common sight, harvesting information every-day that is beamed down to Earth from hundreds of orbiting television and communication satellites . To the casual observer it may appear that we have already conquered space. There is some truth in this. We already have the technology to get to orbit and then go places once we get there.

So it is reasonable to ask if we really need something new.

The problem is that much of the technology in use today is the same as it was when modern rocketry began. The Space Shuttle and every rocket on the launch pad today use the same chemical propulsion concepts as the first rockets of 80 years ago. However, just because technology is old doesn't mean it's outdated. Telephone lines, for example, are old but they still work for the Internet. The combustion engine in your car is the same basic concept as the Model T and works reliably for millions of people every day. Old technology is fine as long as it will meet future needs. As the Internet grows, users will want ever-faster connections and telephone lines and cable modems will be insufficient. At that point new technologies must be developed. As fossil fuel reserves continue to be depleted and costs rise, new energy technologies will be needed to meet the demand. Current rocket technology is at this point today. To go further, something new is needed.

In rockets today, chemicals are combusted to get the thrust a rocket needs to leave the ground and get to orbit. Chemical propulsion, as it is today, is reaching its full potential. Rocket designers are working to refine and improve existing chemical

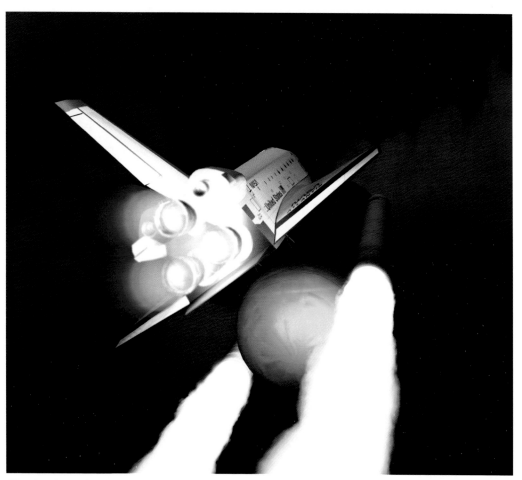

The basic technology on the space shuttle today is over 20 years old.

The History of Rocketry

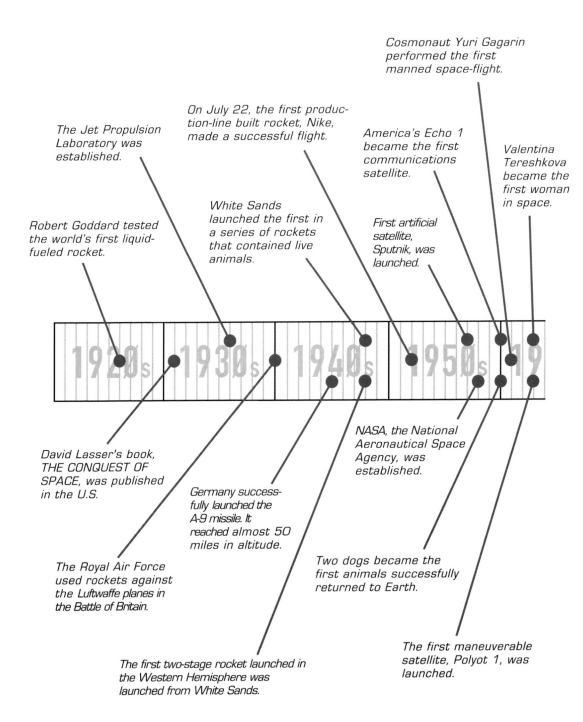

Cosmonaut Yuri Gagarin performed the first manned space-flight.

The Jet Propulsion Laboratory was established.

On July 22, the first production-line built rocket, Nike, made a successful flight.

America's Echo 1 became the first communications satellite.

Valentina Tereshkova became the first woman in space.

White Sands launched the first in a series of rockets that contained live animals.

Robert Goddard tested the world's first liquid-fueled rocket.

First artificial satellite, Sputnik, was launched.

David Lasser's book, THE CONQUEST OF SPACE, was published in the U.S.

NASA, the National Aeronautical Space Agency, was established.

Germany successfully launched the A-9 missile. It reached almost 50 miles in altitude.

The Royal Air Force used rockets against the Luftwaffe planes in the Battle of Britain.

Two dogs became the first animals successfully returned to Earth.

The first two-stage rocket launched in the Western Hemisphere was launched from White Sands.

The first maneuverable satellite, Polyot 1, was launched.

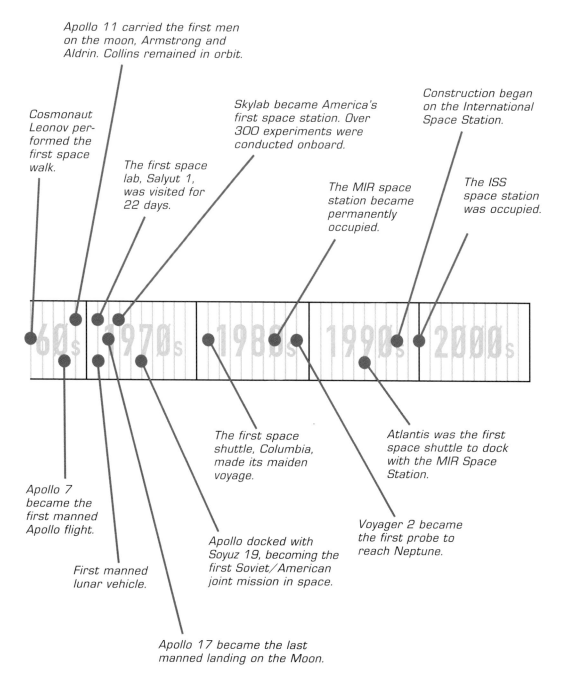

Apollo 11 carried the first men on the moon, Armstrong and Aldrin. Collins remained in orbit.

Cosmonaut Leonov performed the first space walk.

Skylab became America's first space station. Over 300 experiments were conducted onboard.

Construction began on the International Space Station.

The first space lab, Salyut 1, was visited for 22 days.

The MIR space station became permanently occupied.

The ISS space station was occupied.

The first space shuttle, Columbia, made its maiden voyage.

Atlantis was the first space shuttle to dock with the MIR Space Station.

Apollo 7 became the first manned Apollo flight.

Voyager 2 became the first probe to reach Neptune.

First manned lunar vehicle.

Apollo docked with Soyuz 19, becoming the first Soviet/American joint mission in space.

Apollo 17 became the last manned landing on the Moon.

technologies but only marginal improvement in performance is expected. However, chemical propulsion is not dead by any means. In fact, for the foreseeable future, chemical propulsion will remain the primary, if not the only propulsion option for Earth to Orbit (ETO) and for most in-space missions. NASA and the space industry will continue to improve the technology and push the designs as hard as possible until the next breakthrough is found. They will improve and refine, squeezing every drop of performance with better nozzles, fuels, and more efficient designs. Chemical propulsion is being improved but better designs and more efficiency will not deliver the energy needed to take the next major steps in space flight.

So what is the next propulsion solution? That really is the big question. There are many answers and opinions, which often come with a great deal of discussion and controversy. This book will steer clear of the controversy. Many well meaning and respected scientists disagree about what should be done. Many even disagree about what can be done. After working in this industry for years, I have come to understand that there is a fine line between fact and opinion. I have yet to be able to determine where that line is and am unwilling to draw it for myself or anyone else. For that reason, this book will simply present what is on the table without regard to feasibility or controversy. I'll leave the arguing to the innovators, scientists, and politicians.

The Economic Barrier

Today's propulsion technology is expensive and risky. This combination creates a very real barrier to progress in space. Many ideas and technologies for space never get to the launch pad because the cost is just too great. In order to bring a space technology to reality, the inventor must not only create the technology, but also pay the high cost of space delivery. It is much like inland boat builders who must build their boat, then ship it to the water. In the case of space, the shipment cost can exceed the cost of building the product itself.

To illustrate this point, imagine that you have invented a fantastic machine. Your new invention will convert raw lead into pure gold! You are sure to be rich but there is only one catch. Your invention only works in the microgravity of space. No problem, you will simply put raw lead on a rocket,

Ground Shipping Costs vs. Earth to Orbit Costs:

		Typical Cost	Cost in Space
	Gallon of Milk	$3.50	$70,003.50
	Flashlight	$10.00	$5,010.00
	Laptop Computer	$2,000.00	$52,000.00
	27" Television	$350.00	$400,350.00

Cost of Access to Space:

		Delivery Cost	Space Delivery
	Business Letter	$0.37	$625.00
	Overnight Letter Pack	$13.00	$10,000.00

* Based on published rates by weight at the time of this writing.

send it up, and ship the gold back —
you're all set. Unfortunately, your
invention is just not worthy of space.
While you might think turning lead
into gold would be profitable, you
would go bankrupt ferrying the lead
into space. The shipping cost to carry
the lead into space is actually more
expensive than the value of the gold
you bring back. You will need a better
invention if you want to make money
in space with today's launch technology.

Now imagine how the typical
business on Earth would be impacted
if shipping costs were just ten times
higher than they are today. Imagine
licking a $3.70 stamp to pay a bill.
Imagine paying Federal Express $130
to deliver a letter pack. Imagine a roll
of 100 stamps costing $370. This
would make it hard to conduct business,
as we know it. Increased shipping
costs affect the price of everything.

The shipping of parts would be so
high that many cheaper products
could not survive in the marketplace.
In the face of this problem there
would no doubt be a flurry of activity
to reduce shipping costs so business
could function. This is much the
same problem with space technology.
In order to harvest the true benefits
of space we need better and cheaper
space shipping technology.

Now imagine how the inverse
would work. Today shipping costs to
space are about $10,000 per pound.
If the cost could be lowered to
$1,000 or even $100, things would
be quite different. Businesses could
afford to think bigger and be more
aggressive. More ideas would be
possible. Reducing the cost of
access to space would revolutionize
the space frontier.

What Spaceflight Really Costs You Today

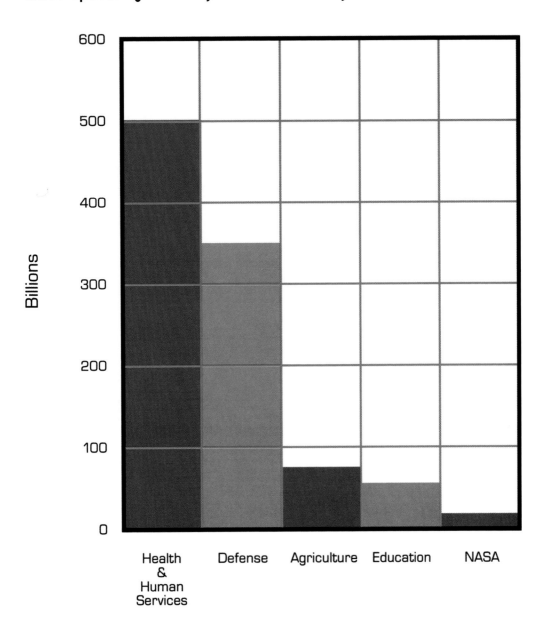

NASA's 2003 budget is $14.6 billion or less than 1% of the Government's total annual budget. That comes to about $50.29 for each person in the United States per year. This is less than the cost of dinner and a movie for a family of three. NASA's budget has remained consistent and only a 2% average growth in funding is anticipated between 1998 through 2004.

The best of prophets
of the future is the past.

— Lord Byron

Chapter III
Looking Back to the Future

Chapter III
Looking Back to the Future

Anytime we look to the future we should look to history as an example. In the case of space transportation we can look back to other types of transportation to see what effect these advancements have had on our lives. History provides some interesting perspectives.

Ocean Exploration — Pushing boundaries and opening new markets

Christopher Columbus successfully found land on his fourth and perhaps last voyage to the New World. He wasn't looking for a new world of course. He was looking for a safer, cheaper route to the Orient. In those days it was a long and dangerous trip from Europe to the Orient. The successful journey could be quite rewarding, as spices and other rare products brought a high price back home. As a result, nobility spent a significant amount of money to discover a new trade route. The first expeditions failed, but step-by-step, the explorers learned more and were eventually successful.

As the early explorers faced the ocean they had great obstacles to overcome. The largest obstacle was the unknown. They did not know what lay over the horizon, where they could get supplies, what the seas might be like, how the ocean currents ran, how hostile the natives might be, or how long the journey would take. The odds of success were definitely not in their favor. Columbus set sail across the Atlantic because he thought India lay over the horizon. Vasco De Gamma and others chose to venture around the tip of Africa. Many explorers had failed in the past but with each failure more was learned about what was needed to

be successful. Ship hulls were made deeper to hold more supplies. Sails were adjusted to give more speed and control. Rudder technology was improved for better control. Maps were improved and ports plotted so they could get supplies. These incremental steps lead to final success.

The early explorers had more than riches to motivate them. There was a very strong motivation to promote their nations and religious beliefs. The nation that dominated the trade routes would have a clear advantage over others. In the process, the Church would spread its religious doctrine. In this regard, technological advancement promoted commerce, nationalism, and religion.

There are some very strong similarities of early ocean expeditions to space exploration. The space race of the 1950s and 1960s between the Soviet Union and the U.S. was about technological and ideological leadership. Each nation wanted to be the leader in space. When the Soviet Union launched Sputnik, the entire world watched with awe. Sputnik proclaimed to the world that the Soviet Union was the first in space and the world leader in space technology. The U.S. was not far behind, but it was behind. The Soviet Union led the way in space until the Eagle landed on the Moon. The day an American stepped on the Moon the United States became the leader in space. The United States took the lead and arguably dominated space exploration in the 1960's but that dominance has waned as other countries have began to reach toward the promise of space.

Ground Transportation — Making personal transportation possible

Less than 100 years ago you would have been one of two types of people. You either lived in the city or on the farm. You lived where you worked. If you lived out in the countryside you worked the land, animals, or supported that activity in some way. If you lived in the city you worked in a factory or some service related industry. The invention of the automobile changed this way of life. When the automobile became affordable to the average person, he had the choice of living where he wanted and working elsewhere. The affordable automobile meant freedom and helped define the 20th century.

The automobile did not start out as an affordable vehicle. It took years of innovation before the masses could benefit from personal transportation. Initially the auto industry assembled automobiles piece-by-piece and each worker took parts to the vehicle to assemble it. This took a significant amount of time, slowed output and increased cost. This meant that only the "well-to-do" people could afford an automobile. Automobile pioneers looked for a better way to manufacture the automobile and reduce its cost. After a fire destroyed his factory, Ransom Eli Olds came up with a new idea where the parts were manufactured off site and brought to the factory. This meant that the worker had all of his tools and parts at hand and ready for assembly. This increased production dramatically from only 425 cars in 1901 to 5000 cars in 1903. Naturally the other auto manufacturers realized the benefit of the assembly line and followed his lead.

Henry Ford improved the assembly line and further improved manufacturing processes. His goal was to make the automobile affordable. His first Model T rolled off the assembly line in 1908 with a price tag of $850. This was still out of reach to the common family. Most families could neither afford nor justify such expenditure for a luxury and novel item as an automobile. In 1913, The Ford Motor Company installed a motorized assembly line to accelerate production. This reduced the time to make a Model T from 12 ½ hours to little more than

Consider how the automobile has changed your life. Could you do without this technology? How is your life altered when you do not have personal transportation? For those who live in inner cities this may not be as inconvenient but for a large part of the population it is a significant problem when the family car breaks down. Imagine how your education, recreation, and career would be changed if the automobile was only available to the most wealthy.

1½ hours. Within 3 years Ford had reduced the cost of the Model T to just $400. Between 1908 and 1927, Henry Ford sold more than 15 million Model Ts to the common man, representing more than half of the automobiles sold in the United States during that time.

Air Travel —
A Global Industry

The airplane experienced tremendous growth in the years after the Wright brothers. The 1930's brought significant engineering advancements. Technological investments and advancements paved the way for bigger and bolder plans. Streamlining of the wings and fuselage allowed the airplanes to cut through the air more efficiently and to fly higher. As the airplanes flew higher, pilots and passengers had difficulty breathing as the air grew thinner. Pressurized cabins were developed to allow aircraft to climb to greater heights. Propellers were designed for adjustment in blade pitch during flight to improve speed and efficiency. Improved radios, autopilots, and gyropilots enabled aircraft to be confidently flown at night and in bad weather.

The 1930's provided the springboard for the jet age. The military provided the rationale and investment needed to develop and fly the next generation of aircraft — the jet. In 1939 German engineers built the first successful jet aircraft. During the 1940's, engineers worked to improve jet performance. Almost 50 years after the Wright brothers' first flight, the first jet airliner, the De Havilland Comet began commercial service. The Comet flew at 500 miles per hour and a high altitude. However, the Comet had engineering flaws that resulted in several fatal accidents. The cabin was pressurized for the high altitudes but as the air outside grew thinner so did the pressure pushing against the outer walls. As a result of the cabin exploding during flight, stronger cabins were developed.

Jet propulsion matured and in 1958 Boeing introduced the 707. Twelve years later Boeing launched the first jumbo jet, the 747, that carried 500 passengers. The first 100 years of powered flight have witnessed fantastic growth and technological advancement. The world is more accessible to more people and global commerce is one of the many byproducts of aviation. Advances in aviation have changed the world. It is difficult to imagine how the world would be today without air travel.

The 1903 Wright Flyer was the world's first powered aeroplane.

New and improved modes of transportation have changed our lives. Technology allowed ancient sailors to stretch the boundaries and go farther with each step. Consistent investments in time, energy, and money led to eventual success. Automobile manufacturers developed new processes to make their product cheaper and more reliable. The airplane industry has made the world smaller and created possibilities and markets that never existed before. It is hard to imagine what our world would be like without any one of these revolutionary movements in transportation history. Today we must decide if we are to realize the true potential of space transportation within the next generation. We have only begun to explore the new frontier of space. The space frontier will belong to those who decide to go and explore.

Milestones in the First 100 Years of Powered Flight

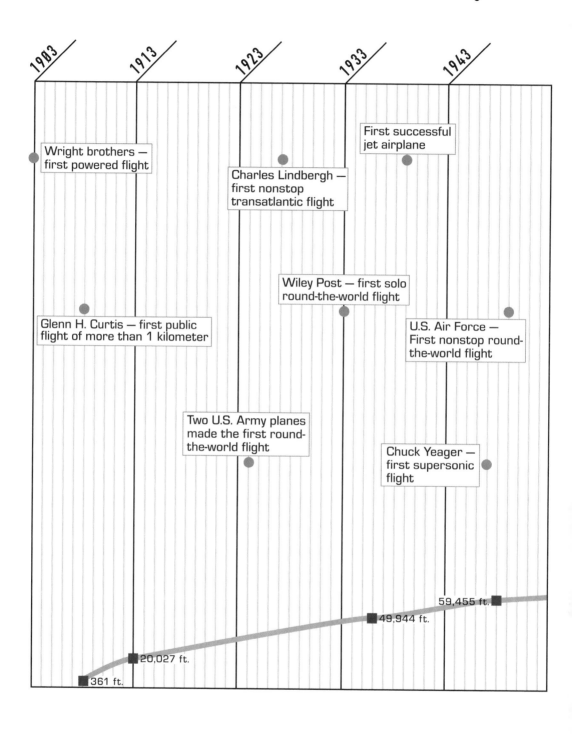

1903   1913   1923   1933   1943

First successful jet airplane

Wright brothers — first powered flight

Charles Lindbergh — first nonstop transatlantic flight

Wiley Post — first solo round-the-world flight

Glenn H. Curtis — first public flight of more than 1 kilometer

U.S. Air Force — First nonstop round-the-world flight

Two U.S. Army planes made the first round-the-world flight

Chuck Yeager — first supersonic flight

59,455 ft.

49,944 ft.

20,027 ft.

361 ft.

● Events in Aviation History   ■ Altitude Records

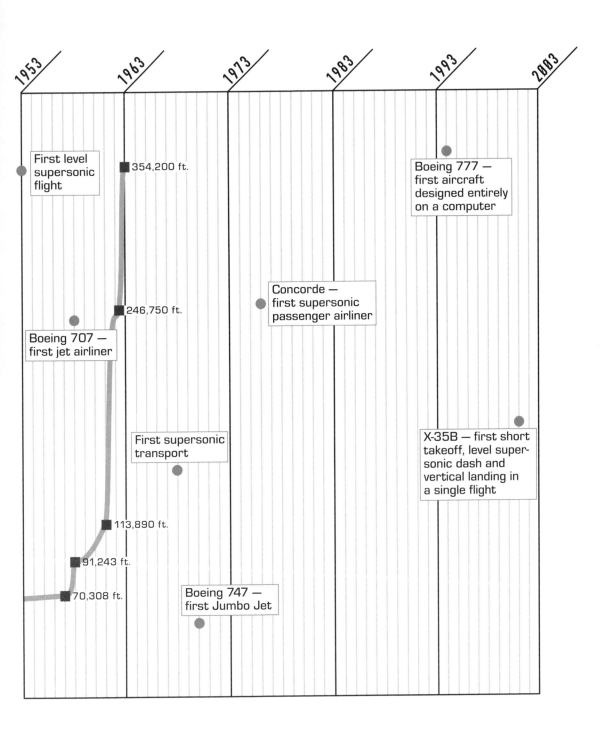

1953 | 1963 | 1973 | 1983 | 1993 | 2003

First level supersonic flight

354,200 ft.

Boeing 777 — first aircraft designed entirely on a computer

246,750 ft.

Boeing 707 — first jet airliner

Concorde — first supersonic passenger airliner

First supersonic transport

X-35B — first short takeoff, level super-sonic dash and vertical landing in a single flight

113,890 ft.

91,243 ft.

70,308 ft.

Boeing 747 — first Jumbo Jet

You must not have too much fear
of not being up to your task
when you are approaching
great problems and great works.

— Georges Duhamel

Chapter IV
Earth to Orbit

Chapter IV
Earth to Orbit

One of the most interesting aspects challenging the engineers and scientists trying to get into orbit is the fact that they really don't have that far to go. The edge of space is a relatively short trip. To get into LEO you have to go about as far as from New York City to Washington D.C. It's about as far as Chicago is from Detroit, London to Paris, and less than the distance from Los Angeles to San Francisco. Its only about 200 miles. If you had a road to space, you could drive the family car there in a few hours — ignoring the obvious limitations of course.

While the distance to orbit is not significant, the speed required to get there and stay is substantial. A satellite in orbit is being pulled by Earth's gravity and is actually falling downward but it is high enough and moving fast enough that it misses the ground as it falls. The lower the orbit is, the stronger the gravitational pull and the greater the speed required for a satellite to stay up. At 200 miles above the Earth, a satellite must be traveling at more than 17,000 miles per hour to stay up. A higher orbit of 22,000 miles would require a speed of only 6,900 miles per hour.

Gravity is strongest at the surface of the Earth and weakens as you go up. A rocket sitting on the launch pad has the difficult challenge of lifting off the ground where gravity is the strongest and its fuel tanks are at their fullest. The speed of the rocket depends upon the mission. A mission to the Moon would need to reach a speed of over 24,000 miles per hour, for example, to escape Earth's gravity altogether. Speed, not distance, is the challenge for Earth To Orbit (ETO) vehicles.

Thermosphere
(50 – 185 miles)

Mesosphere
(30 – 50 miles)

Stratosphere
(12 – 50 miles)

Troposphere
(0 – 7 miles)

Space Shuttle Upgrades

The space shuttle is a marvelous machine. It is the first and only reusable spacecraft ever made and deployed. The space shuttle is built using 1970's technology and is very expensive to operate, as much as $500 million per flight. The shuttle budget has historically consumed a sizable portion of NASA's budget. The shuttle is expensive because after each flight it must be refurbished to be flight ready for the next trip. The entire vehicle must be inspected largely by hand. Any damaged or expended components must be replaced or repaired. This process takes a great deal of manpower and is very expensive. It is so extensive and expensive that some have said the shuttle is actually more "salvageable" than "reusable."

NASA does not presently have plans to replace the shuttle, but to extend its life. In order to extend the life and improve the reliability and safety of the shuttle, the following upgrades are being considered:

- Advanced Health Management System for the shuttle main engines
- Electric auxiliary power unit to eliminate hazardous fuel and high-speed equipment to power the hydraulic system
- Enhanced orbiter avionics for better situational awareness
- Improved main engine operating environment
- New solid rocket booster auxiliary power unit to eliminate hazardous fuels used to control thrust direction
- Improved solid rocket booster hold-down hardware to reduce chances for human error and improve design
- Improve solid rocket booster propellant to increase performance
- External tank friction stir welding to improve manufacturing control and improve reliability

On February 1, 2003, the Space Shuttle Columbia broke apart during reentry. Columbia was NASA's first orbiter and reached orbit for the first time in 1981. Twenty-two years later, Columbia was still in use. Columbia had flown 28 missions. Columbia had extensive technology upgrades and all shuttle upgrades were originally scheduled to be completed by 2005. At this time it is unknown if they would have changed Columbia's fate. This event will no doubt renew interest in space transportation for both supporters and detractors alike. This event will test our commitment to space transportation and America's commitment to future leadership in space.

Second Generation
Launch Vehicles

Technological advancement is incremental and happens one step at a time. The next step along the technology road for Earth to orbit is the development of a Reusable Launch Vehicle (RLV) that is safe, reliable, and affordable to operate. The shuttle has three major components: the orbiter, external tank, and solid boosters. The orbiter and solid boosters are reused while the external tank is expended. A second generation reusable launch vehicle would be like the external tank and solid boosters but be fully reusable. While NASA is not working on replacing the shuttle in a single effort at this time, simultaneous efforts are underway to develop a Crew Rescue/Transport Vehicle (see Orbital Space Plane) and develop the technology for a potential second and third generation RLV (see "Next Generation Launch Technologies"). Many ideas have been proposed and the designs are quite varied. The following are some of the ideas and images of second generation RLVs.

Typical Second Generation Mission Profile

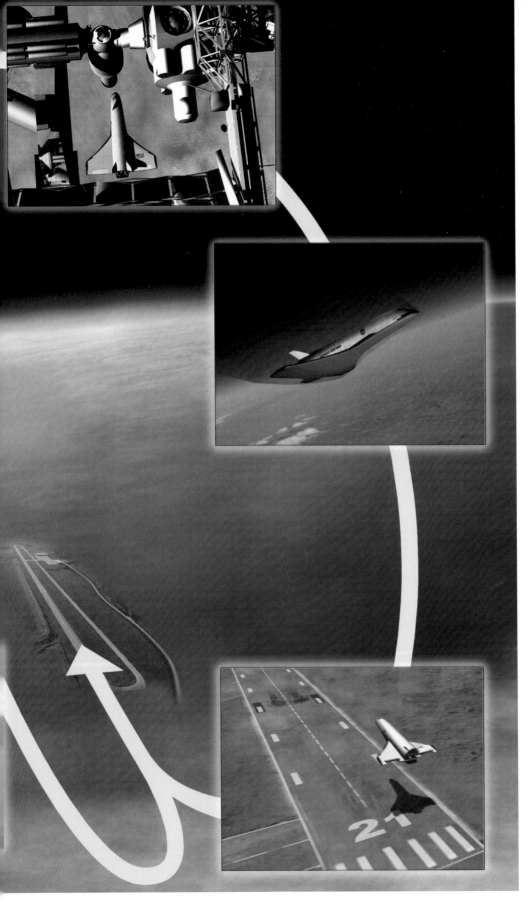

Air-Breathing Combined Cycle

I think we all have seen commercials on television where a new appliance is being demonstrated to "slice, dice, and even make julienne fries." These kitchen wonders claim to make short work of almost everything. I have one in my kitchen cabinet some place. A more useful example of these multifunction machines is the all-in-one computer printer, scanner, copier, and fax machine. If you need all of these functions, this is an economic choice. It is much cheaper to buy this single machine rather than buy each appliance separately.

Another example of a multifunction machine is the transmission in a truck. The transmission is designed to help the truck get to its destination and adapt to the various conditions throughout the trip. The transmission has low gears for startup and steep hills. High gears are used when the truck is on the open road. Without the ability to change gears, the performance of the truck would be

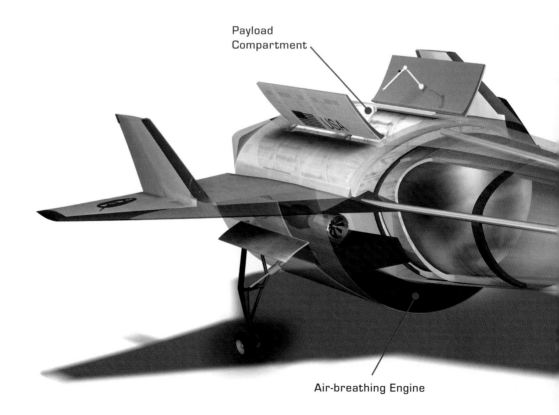

Payload Compartment

Air-breathing Engine

significantly limited. With a single gear the truck would be efficient in one driving mode but challenged in all others. If geared for the open road it would have difficulty getting up hills and on take off. If geared for hills, the engine would have to work extra hard to maintain the speed limit on the highway. The multiple modes of the transmission help the truck perform better throughout the trip.

An Earth To Orbit or ETO mission has unique challenges. The vehicle must leave the launch pad, accelerate, and reach orbital velocity. In the process, the craft moves through different atmospheric conditions and pressures. Gravity is the greatest on the launch pad and virtually non-existent in orbit. Vehicle skin temperatures rise as the vehicle goes faster. This makes a single engine design extremely difficult. Just like the truck, the spacecraft would benefit from "shifting gears" — a mode for launch, a mode to get up to speed, a mode for high

Typical Components of an RBCC Vehicle

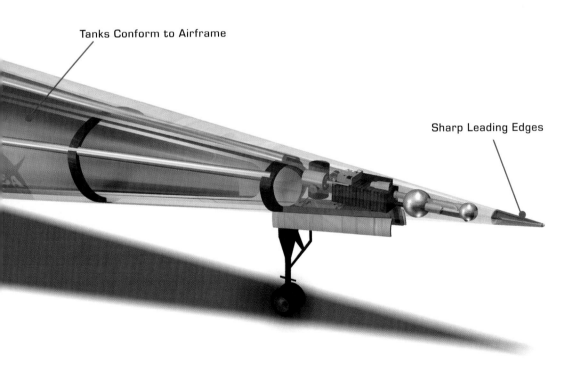

Tanks Conform to Airframe

Sharp Leading Edges

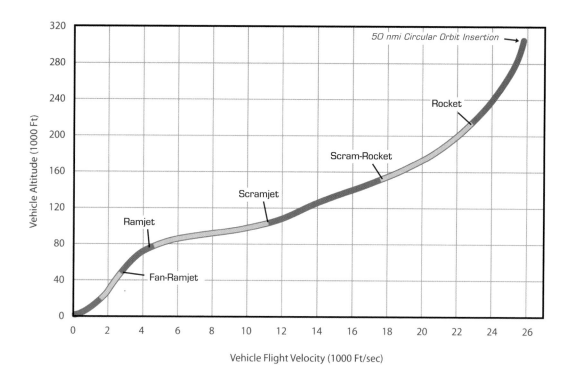

Vehicle Altitude (1000 Ft)

50 nmi Circular Orbit Insertion

Rocket

Scram-Rocket

Scramjet

Ramjet

Fan-Ramjet

Vehicle Flight Velocity (1000 Ft/sec)

altitudes, and a mode to reach orbit velocity. Unfortunately, there is no transmission equivalent for rocket engines today that will allow it to switch modes. Back in the 1960's, engineers envisioned a combined rocket engine that would merge various air-breathing propulsion technologies into a single spacecraft. Like the "all-in-one machines," a combined spacecraft engine could benefit from having different engine technologies and operating modes that could be shifted through as flight conditions change.

There are many names and types for combined cycle vehicles: Rocket Based Combined Cycle (RBCC), Turbine Based Combined Cycle (TBCC), Air Augmented Rocket (AAR), and

Air-breathers. A combined cycle engine is an all-in-one engine that combines a jet engine, ramjet, scramjet, and rocket engine on a single spacecraft. The idea is relatively simple. The spacecraft will use each engine technology just like gears in a transmission. When the engine type has reached its maximum benefit it is shut down and a new engine type will shift in.

A rocket engine combines propellant and an oxidizer with a spark to create a combustion that pushes against the engine nozzle to propel the rocket. When the Space Shuttle is sitting on the launch pad, a large portion of the weight of the vehicle is oxidizer (normally Liquid Oxygen, or

LOX). Oxidizer is pumped into the rocket engine along with fuel where it combusts and propels the vehicle upward. The air that the spectators breathe near the launch pad is an excellent oxidizer because it contains oxygen. A rocket engine pumps the oxidizer into the combustion chamber at launch and throughout the flight because as it climbs into space, the air gets thinner and the engine would stop burning if it were "air-breathing." Sitting on the launch pad or runway, and in the early stages of flight, a combined cycle engine would use the outside air as an oxidizer source rather than from onboard fuel tanks which it uses at extremely high altitudes and in space. This means the combined cycle launch vehicle is much lighter and potentially more efficient.

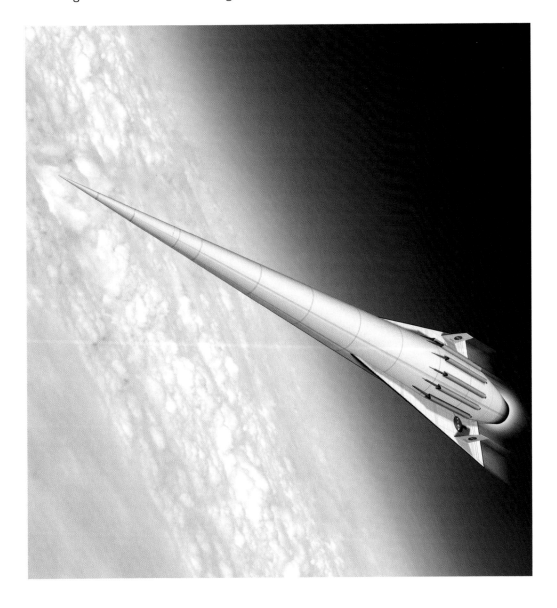

There are many configurations and variations of air breathing launch vehicles. Some designs have separate engines for each mode. Some concepts are Single Stage To Orbit (SSTO) and some are staged. To understand the basic idea, I will walk through an example flight and describe what is happening in the engine at each phase.

Sitting on the launch pad or runway, the crew turns the engine on in fan mode. In fan mode, the engine brings air in, compresses it, and exhausts it at high speed out the back of the vehicle. A turbine is used to accelerate and compress the air much like a commercial jet engine. This is the "low gear" for the vehicle that doesn't use much fuel but has relatively low thrust. Afterburners are ignited at the back of the engine to get an additional push. Internal

X-43B in flight.

rocket engines are ignited to give the thrust needed for takeoff. This is not your average "low gear" however, as this mode will carry the spacecraft 30,000 feet at a speed between Mach 2 to 3+ (Mach 1 is the speed of sound). At this point the spacecraft is working much like the typical supersonic airplane.

To go faster the spacecraft needs to change gears and shifts into the straight air-breathing ramjet mode. Ramjet engines have been successfully demonstrated and work only when the vehicle has gotten past Mach 3. At Mach 3, the vehicle is moving fast enough to compress the airflow into the engine by its sheer speed alone. The inlet design of the engine lets the right amount of air in and the air is compressed in the process. As the air gets compressed it heats up, really heats up. The afterburner now serves as a "ramburner" and provides the push needed to take the spacecraft to 90,000 feet and Mach 7. The ramjet engine could do more, but the temperatures will continue to rise and other engines can handle this situation better. One such engine is a scramjet.

The scramjet engine is currently being investigated but it has not been extensively flight-tested for a variety of reasons. Since the scramjet must

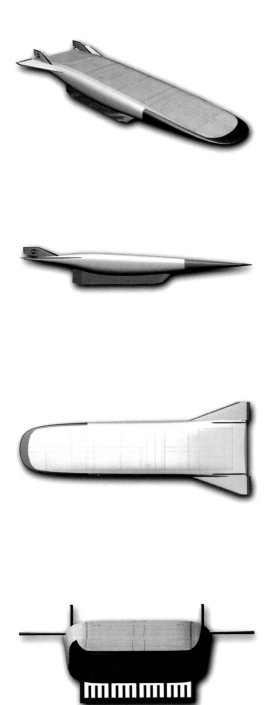

X-43B — TBCC version.

operate at supersonic speeds above Mach 7, it is not suitable for wind tunnels and must be flight-tested to truly evaluate the design. The scramjet opens the inlet to let more air into the engine. The fuel injectors at the rear of the engine are turned off and injectors at the center of the engine are turned on. With the injectors at the center, there is more time for the fuel to combust inside the engine. At speeds between Mach 7 and 15, fuel injected into the airflow at the rear of the flow would just blow out the back and no benefit would be had from the combustion. With the fuel injectors positioned earlier in the flow, the fuel is injected into the super-hot air, combusted, and theoretically accelerates the vehicle before it exits the engine. Scramjet mode will take the spacecraft to Mach 15 and about 130,000 feet.

At 130,000 feet the air is getting extremely thin so the rocket engine fires and adds a little punch to push the vehicle beyond Mach 15. At around 160,000 feet there is insufficient oxygen to support the air breathing engines, so the outside inlet is closed and the vehicle runs on pure rocket power. Propellant and oxidizer from the onboard fuel tanks feed the rocket engine. In rocket mode, the spacecraft can reach 300,000 feet and a speed of Mach 25.

When you look at air-breathing rockets you will notice that many of them look quite different than historical rockets. Air-breathing rockets use the overall vehicle shape as a component of the propulsion system. The vehicle body, or airframe, works to help compress and direct the airflow as it reaches the engine. This is one of the reasons that there is so much variety in the way these vehicles look.

Synerjet

Synerjet is one of the original concepts from the 1960's. It features a series of engines around a conical body. The conical body might also have grooves or ducts that would channel the airflow into the engine. Since the vehicle had multiple engines, it could be serviced more easily and an engine could be quickly replaced if the need arose. Synerjet would take off vertically like a traditional rocket and land in the same orientation. The engines would transition through the various modes as described earlier. At landing, the engines would fire to slow the vehicle and deploy retractable legs. The spacecraft would then land in the same vertical orientation as take off using a fuel-thrifty fan mode.

Synerjet would return bottom first then flip nose down for the approach to the landing area.

Hyperion

Georgia Tech University is known for its unique contribution in aero-space. Hyperion is a concept that grew out of the school and embodies many of the challenges facing the next generation of reusable launch vehicles. It has conformal fuel tanks. Traditionally, fuel tanks have been spheres or capsule shaped. Some designers believe that it is important to develop irregular shaped tanks to allow the systems engineers more flexibility in design. Hyperion also has a sharp leading edge, which will require advanced thermal protection. Hyperion would take off and land on a conventional runway.

*Hyperion would take off from
a conventional runway.*

A Vision of Future Space Transportation   53

Argus

Like Hyperion, Argus is a product of Georgia Tech. Argus has two external engines. One of its most notable features is that it relies upon some sort of launch assistance. Some research has been conducted to evaluate how electromagnetic catapults could be used to help vehicles like Argus get off the ground and on their way into space. So far no cost effective launch assist system has been developed.

Argus would land on a conventional runway.

Spaceliner 100

Spaceliner is something of a derivative of Argus. One limiting factor with Spaceliner 100's design is engine placement. The engines are out from the fuselage, where Argus had them close in. By being close to the fuselage the air is somewhat compressed as it enters the engine. Spaceliner 100 loses this advantage. This design anticipated using a magnetic track to levitate and accelerate the vehicle for launch. Spaceliner 100 is not seen much these days, but was a featured design for a time.

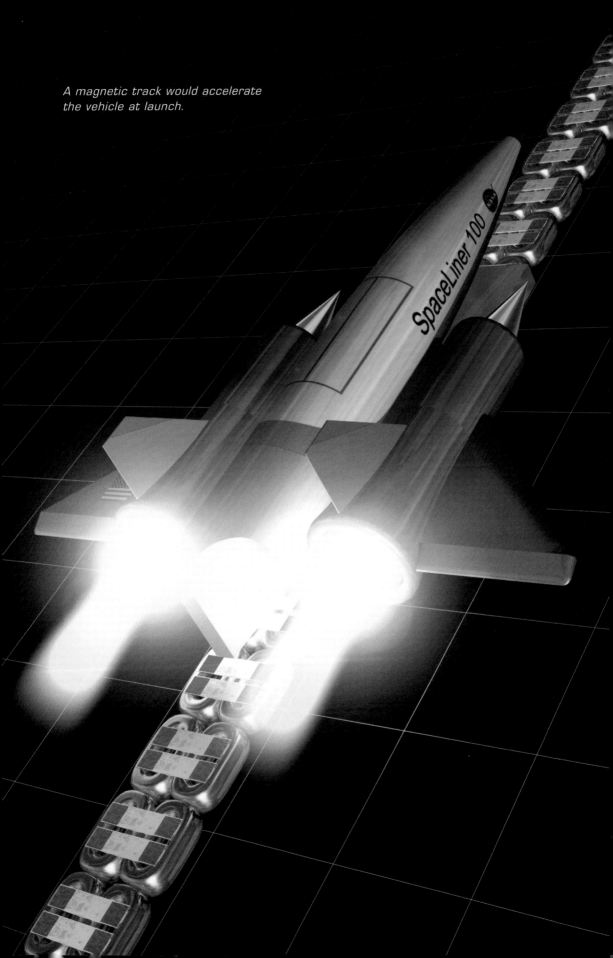

A magnetic track would accelerate
the vehicle at launch.

D-21

The D-21 was an experimental vehicle from the Cold War era that would take pictures at a high altitude and speed. The engine was a kerosene-fed Ramjet that would reach Mach 3. The D-21 was successfully launched from NASA's high-altitude SR-71 and flew a couple of test flights but was plagued by mishaps that ended the development effort. Some scientists have considered bringing the D-21 out of retirement and using it as a test vehicle for new engine studies.

The underside of the airframe in front of the engine helps direct airflow into the engine.

Air Breathing Launch Vehicle (ABLV)

The ABLV concept has several variations. Each ABLV design variant is numbered. The ABLV-7 is pictured here. The design has several notable characteristics. The engine is underneath the vehicle and is divided into compartments. Each compartment can be a different engine mode. Instead of combining all of the modes into a single inlet and air flow-path, the ABLV would turn one chamber on and the others off depending upon the flight conditions. The curved underside of the front helps funnel compressed air to the engine. The basic shape of the ABLV appears on numerous designs.

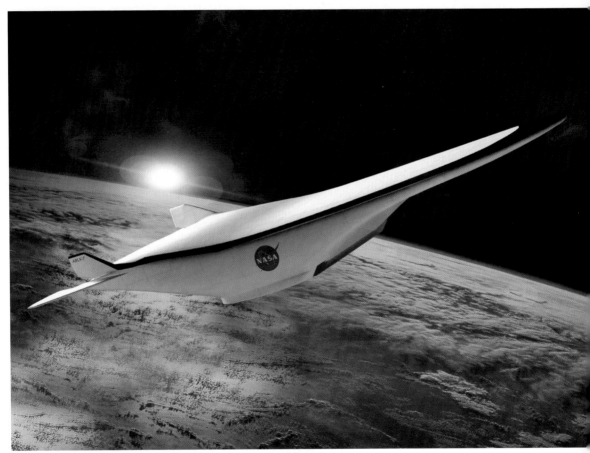

Numerous versions of the ABLV have been considered.

Two-Stage-To-Orbit Designs

A popular concept is using the combined air breathing rocket engine as a reusable first stage and a small rocket powered orbiter as the second. The orbiter would ride piggyback up to a designated altitude and be launched into orbit.

Configurations vary but staging would occur suborbitally and at a speed between Mach 8 to 10. The first stage would glide back and land on a runway. Since the first stage would not go into orbit it would need only minimal thermal protection.

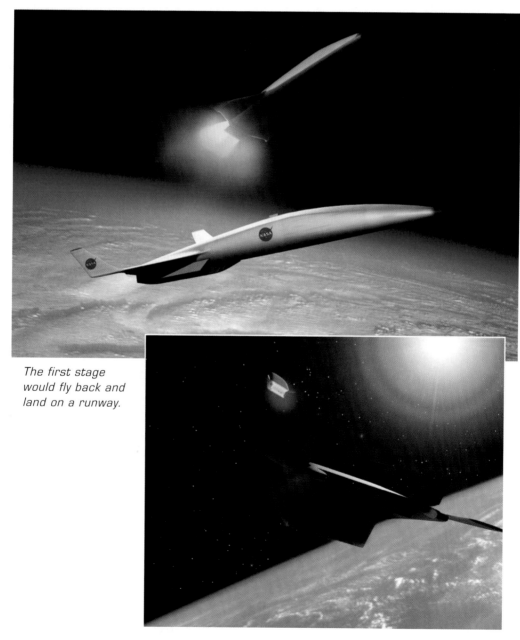

The first stage would fly back and land on a runway.

Staging occurs at the edge of space.

The Astrox design features an inward turning engine.

Astrox

Astrox Corporation has an RBCC concept that has taken the idea of conforming the airframe to the engine to a whole new level. The Astrox concept features an inward turning design where the whole front end of the spacecraft directs air into the air-breathing engine. The Astrox vehicle looks like an engine with wings. The design has the advantage of compressing a large volume of air as it enters the engine. Astrox will have to address some technical challenges. Like Hyperion, Astrox would need uniquely shaped fuel tanks. Tanks may actually have to be doughnut shaped. Astrox would take off and land on a conventional runway.

FASST

This Boeing concept features a two-stage vehicle. The first stage is a turbojet-powered, large wave rider that detaches and glides back to the landing strip. The second stage is powered by a hydrogen-fueled combined-cycle engine and goes into orbit.

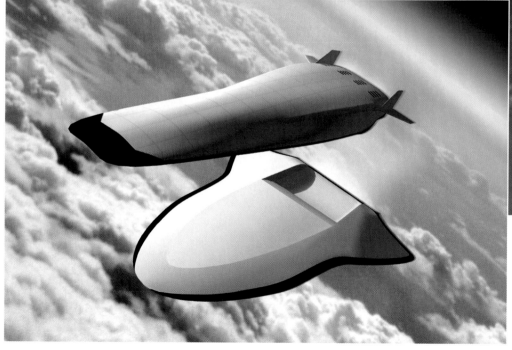

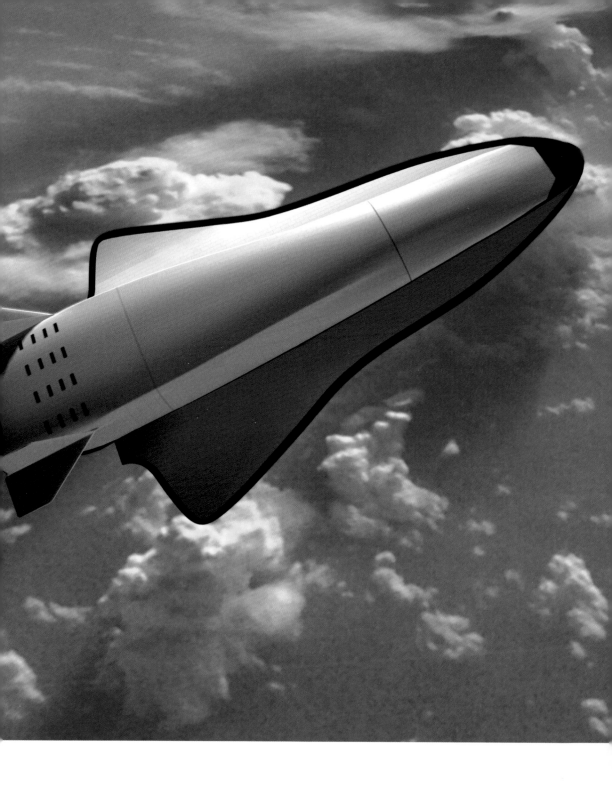

X-43

The X-43 is a flight demonstrator vehicle series. The X-43 will be dropped from NASA's B-52 and flight tested over the Pacific Ocean. The goal of the program is to evaluate the high Mach performance and technology of air breathing propulsion. The vehicles are scale demonstrators that are much smaller than actual spacecraft. The X-43 vehicles will not go into orbit and are only intended to help ready the technology to take the next development steps.

The X-43B has two variations. One is a Rocket-Based Combined Cycle (RBCC) and the other is a Turbine-Based Combined Cycle (TBCC). The RBCC will work as described earlier. The TBCC will be different in that it will not have the rocket component — it will rely upon turbines (fans) for its acceleration. The RBCC is easy to recognize by its frontal fins or horizontal canards. The turbine plus dual-mode ramjet is planned to achieve Mach 7 flight speed.

NASA's B-52 will be the launch platform for the X-43B.

A Vision of Future Space Transportation

The X-43C will be attached to the front of a Pegasus rocket and dropped from NASA's B-52. The Pegasus is a proven launch vehicle developed and operated by Orbital Sciences Corporation. The Pegasus will take the X-43C up to speed where it will start its ramjet engines.

The X-43C will launch from the front of a Pegasus rocket.

ISTAR

ISTAR is not a vehicle but an engine. ISTAR is short for Integrated System Test of an Air-breathing Rocket. This engine is being designed and built by a team of scientists from Aerojet, Rocketdyne, and Pratt & Whitney who are normally competitors in the aerospace industry. ISTAR is the engine planned for flight on the X-43B flight test vehicle.

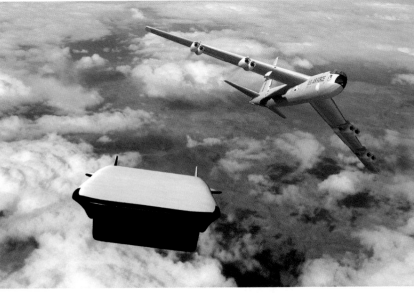

The vehicle will be dropped from NASA's B-52 or similar aircraft.

ISTAR is a part of the X-43B program.

Final Note on Air Breathing Rockets

Research on air breathing rocketry has been going on at some level for 40 years or more. It is a sizable task to bring all of these technologies together into a single vehicle. It has been estimated that it will require around $30 billion of research and development money to have the first usable vehicle. More than double that amount to include all of the support and infrastructure required to operate and fly it over 25 years. Considering the fact that launch requests have been declining over the past several years, some question the relevance of a new launch vehicle right now. The market for launch technology is not encouraging to those who are allocating funds.

Space leadership has changed. The United States and Russia once dominated space. While the United

The X-43C in flight.

Stargazer.

States is the only country in the world with a space shuttle, it is not necessarily the leader in space transportation. In recent years Europe has seized global leadership for commercial space launches. Currently Europe has 40 percent of the launch market, which is more than twice that of the United States. The future of space transportation will belong to those who make the choice to pursue it and make the necessary bold steps to capture it.

The benefit of an all-in-one spacecraft is that you get the synergistic benefits of performance and the economy of each technology. One of the fundamental goals of the air-breathing rocket is to develop a fully reusable spacecraft. The ideal vehicle will operate much like an airliner. It will fly a mission, get checked out, refuel, and fly again. With airline-like operation, the spacecraft will open new opportunities in space and open the door to untold technologies here on Earth. The Air Force Scientific Advisor Board summarized the hypersonic air-breathing program as "at the same crossroads" that supersonic aircraft were 50 years ago. The Board continued: "Would any reasonable person today say that the United States... made a mistake in supporting supersonic R&D?"
— *Aviationnow.com*

ALTO — Air Launch to Orbit

Military aircraft have been refueling in flight for years. As an aircraft gets low on fuel, a tanker aircraft flies in front of and lowers and connects a fuel line. Air Launch is a concept that applies a variation of flight fueling to an ETO spacecraft. A winged spacecraft is mounted to the top of a carrier aircraft such as a Boeing 747. On the runway, the oxidizer tanks of the rocket are empty. The vehicle takes off from a conventional runway and during flight the 747 takes in outside air and refrigerates it into liquid oxygen. The liquid oxygen is collected on the 747 and transferred up to the spacecraft. In a couple of hours, the tanks are filled and the spacecraft fires its engines and heads for orbit. The spacecraft would perform its mission then return to Earth and land on a conventional runway like the Space Shuttle does today.

ALTO's primary design goal is to increase safety and reduce the cost of ETO missions. For that reason, existing aircraft were initially chosen. Several aircraft have been considered, but in the end, it was realized that the spacecraft might be too unstable if located on the back of an aircraft. The configuration would have a tendency to roll, cost more to operate, and take considerable time to mount the rocket to the aircraft with a crane or other mechanism. The ALTO design study is now in its next phase.

Initial design study. A smaller spacecraft would be required for flight stability.

The carrier
aircraft could
launch ELVs
or reusable
spacecraft.

The second stage would be dropped and its engines would be ignited
to propel the vehicle to space.

A Vision of Future Space Transportation

ALTO — The Next Generation

In the aircraft industry it is a matter of business practice that the manufacturer must sell at least 100 aircraft in order to be remotely profitable. With this in mind, the ALTO design team has envisioned a box-winged aircraft that will serve as an in-flight launch platform and be useful for other purposes as well. The aircraft, dubbed "CRoSSBoW" (Cargo Rocket Space System Box Wing) could be used to transport a host of cargo modules or any heavy lift items for commercial and military purposes. The advantage of this design is that it could carry the spacecraft on the bottom rather than on the back. This configuration is much more stable at take off and during rocket deployment. The boxed wing configuration allows for thinner, longer wings, which will provide more lift, less drag, and greater structural stability.

The ALTO concept has set aside the idea of collecting LOX during flight for now. Since a new heavy lift aircraft is required for other national interests, it no longer seemed necessary to develop the LOX system for rocket launch now. ALTO will gain economically by serving as a fully reusable first stage for medium to heavy payload expendable rockets. This is similar to the Pegasus launch model offered by Orbital Sciences, but would deliver much larger payloads at significantly less operational costs. ALTO would serve as a near term stepping stone to reducing the cost of getting to orbit and help develop and test the next generation of air-breathing rocket concepts. No radically new technology would be required and current expendable launch vehicles (ELV) could be utilized.

Microwave Lightcraft

Flying saucers have been around for a long time, in science fiction that is. Dr. Leik Myrabo is working on a concept that actually looks and flies like a flying saucer. Powerful microwave generators would be focused on the spacecraft and provide power for flight. The greatest challenge facing the Microwave Lightcraft will be in solving the substantial power requirements. For more information about Microwave Lightcraft look for a future book by Dr. Leik Myrabo. The book, planned for release in late 2003 by Apogee Books, will be a flight manual for the Microwave Lightcraft.

Gun Launch

There is a famous antique science fiction film that depicts a large gun that fires a shell to the Moon. Gun launch advocates took this a bit more seriously than the rest of us. Gun launch has at least two variants with the same goal of launching a small payload directly into space. The payload could be a small satellite or a package to the International Space Station. An effective gun launch system would be useful for delivering supplies such as water to the space station. This would be more convenient and less expensive than current alternatives, provided the payload can survive G forces of well over 100.

Sling-A-Tron would be a permanent installation in a remote area and use a mountainside for a launch site.

Sling-A-Tron

If you take a garden hose and snap it up and down quickly it will vibrate down the length of the hose and eject any water left in the hose. The Sling-A-Tron takes this idea many steps further. A coiled barrel would be suspended on platforms that could gyrate the coil rhythmically. This sequenced gyration system would accelerate the payload, which is inside the coil, to a high rate of speed and propel it into orbit. The coils would be rather large and require the entire base to be gyrated or each pedestal gyrated rhythmically. This would be an enormous mechanical machine.

Blastwave would be installed on the side of a remote mountain. Little of the launch system would be reusable.

Blastwave

The Blastwave concept envisions placing a barrel on a hill in a remote area. The barrel would be suspended above the ground and a series of explosive charges would line the barrel. The payload would be placed inside the barrel and the explosives detonated just behind the payload. The shockwave from the explosives would accelerate the payload into orbit. Once used, the barrel would be completely destroyed. It has been suggested that a barrel could be constructed that would be reusable but this is not likely with any materials known today. The strength of the explosions required to send a payload to orbit is far too powerful.

*The greater the difficulty
the more glory in
surmounting it.*

— *Epicurus*

Chapter V
In-Space Transportation

Chapter V
In-Space Transportation

Moving about in space is quite different than transportation on Earth. Virtually every method of transportation we use on Earth is not an option in space. Space does however have some very intriguing alternative mechanisms for transportation. There is no air in space but there is solar wind in the form of radiation currents that flow from the Sun. In the inner solar system there is ample sunlight that can be converted to electricity for propulsion. Then you can always carry some potential energy along like chemicals, nuclear, and so on. Space is a vacuum so even minuscule amounts of thrust can push a craft because there is no drag or friction. Moving in space has its own set of challenges and corresponding opportunities. In-space transportation requires you to think quite differently.

Unlike ETO, in-space transportation must deal with enormous distances. Let's look at how far things are out there. Low Earth Orbit (LEO) starts at about 200 miles above the Earth's surface. This is 1/1250 of the distance to our closest celestial neighbor, the Moon. The Moon is about 250,000 miles away. A trip to our moon is like going from Washington D.C. to San Francisco about 100 times. The Apollo astronauts went to the moon and it took them 3 days to make the trip in one of the largest and most powerful rockets ever built. For the average person's thinking, this is a long, long distance. For the scientists looking to explore our solar system, this is like a trip to the mailbox.

Our solar system revolves around the Sun. The Sun gives us our light and warmth on Earth from a distance of 93,000,000 miles away. This distance is a standard unit of measure in space and is referred to as one Astronomical Unit or AU. This is like a trip from New York to London 27,000 times. At 200 miles per hour it would take 1,274 years to get to the Sun. The Concorde flies from London to New York in about 3 hours and the trip to the Sun at this speed would take about 9 years. The Saturn V that took the astronauts to the Moon

left Earth's gravity at 25,000 miles per hour. It would take 155 days to reach the Sun at that speed. Our Sun is a star next door but is certainly not close by.

A photon of light coming from the Sun travels at 186,000 miles per second and takes a little over 8 minutes to get to Earth. The light that we feel today actually started its journey eight minutes and 20 seconds earlier.

Sunlight travels 11,160,000 miles per minute, 669,600,000 miles per hour and 16,070,400,000 miles in a day. Sunlight can travel to the farthest reaches of our solar system in less than a day. Light can travel 5,865,696,000,000 miles in a year. At that speed it would take over four years to reach the nearest star. Thus we say the nearest star, Alpha Centauri C, is 4.3 light-years away.

Solar System
Closest distances from Earth in millions of miles

Sun	.92.7	Saturn	.792
Mercury	.57	Uranus	.1689
Venus	.26	Neptune	.2695
Mars	.48.6	Pluto	.3565
Jupiter	.389.7		

Earth

Saturn

Neptune

These are staggering numbers and far larger than anything encountered in our everyday lives.

It is difficult to describe how large space is. We have nothing in our experience that compares. The vastness of space stretches the bounds of our imagination. Traveling to Mars may take many months but a trip to Saturn will take years. Destinations beyond our solar system require more time than the careers of the scientists that launch the spacecraft. New in-space propulsion systems are needed to explore our solar system in reasonable timeframes. The following pages describe some concepts to advance in-space transportation. Each concept has a different range and scope. Not all missions are the same and each concept can look dramatically different from the others.

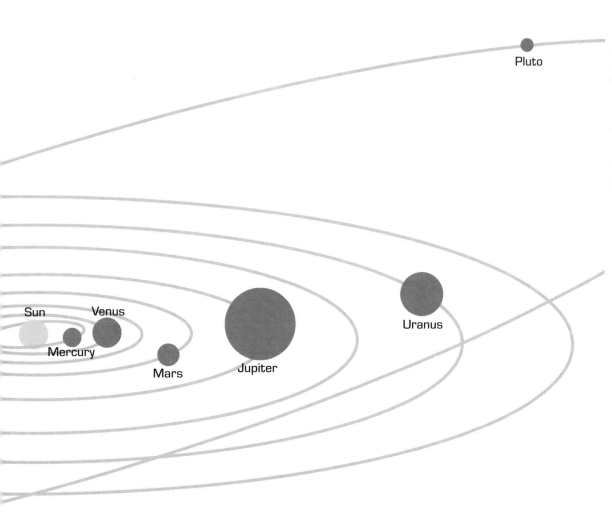

Pluto

Sun

Venus

Mercury

Mars

Jupiter

Uranus

Electro-Dynamic Tether
(ED Tether)

Magnetic forces either attract or reflect. They either pull or push depending upon how they are oriented. If the opposite poles of 2 magnets are placed together, they pull together and when reversed they push apart. In addition, when electricity passes through a metal wire, a magnetic field is produced. An electric motor functions using this basic concept. Magnets are placed in a circle around a wire coil in the center. When electricity is sent to the wire coil it produces a magnetic field that pushes against the outer magnetic field created by the permanent magnets. The motor spins as long as an electric current keeps the magnetic

An ED tether can be used to reboost or deorbit.

field flowing. When the electric current stops, the magnetic field stops, and the motor stops turning. Generators work in the same fashion only in reverse. If the center wire coil spins through the magnetic field, it creates electricity. This is the basic principle that is used to create electricity that powers our homes and businesses only on a very large scale.

The Earth has a significant magnetic field. Electrodynamic, or ED, tethers use this natural magnetic field to get a push for a spacecraft. If a metal wire in a tether is passed through the Earth's magnetic field an electrical voltage is created. If a cathode is placed at one end, current can be created. When the current flows it creates a magnetic field. ED tethers orbit the Earth and pass a metal wire through the magnetic field. The spacecraft is in orbit thus the wire is not stationary, so the movement through the Earth's magnetic field induces a natural electric current in the wire. The magnetic field created by the electric current interacts with the Earth's magnetic field, which results in drag on the spacecraft and slows its orbit. The drag can be used to deorbit dead satellites or spent upper stages. Natural deorbiting is usually a relatively slow process and can take months to years. The upper stages of spacecraft can linger in orbit as space debris for some time. Tethers could be used to quickly deorbit spent space hardware that has served its purpose. ED tethers offer a propellant free propulsion system that will work as long as the Earth has a magnetic field and the space-craft is operational. NASA's ProSEDS mission is scheduled to demonstrate an ED tether deorbit in 2003.

ED tethers can also be used to raise an orbit. If the tether has electricity applied to it as it enters the Earth's magnetic field, it works like the motor. The electrical current creates a magnetic field that is properly oriented with respect to the Earth's magnetic field to repel each other, much like the inside of an electric motor. The tether actually pushes against the Earth's magnetic field to maintain or to raise its own orbit. The ED tether can use solar panels to charge batteries that will supply the electricity. This is especially useful for large spacecraft like the International Space Station that have

Solar panels would power the tether.

to continually fight the small, but continual, aerodynamic drag it experiences in LEO. An ED tether connected to the ISS could continually reboost the orbit without refueling. Currently Russian spacecraft must deliver tons of fuel to keep the space station aloft.

ED tethers are unique in that they offer propellant-less propulsion. They use natural forces to do their job. ED tethers are limited to missions where there is a magnetic field present. The process works well in low orbit but as the orbit gets higher the magnetic field is weaker and ED tethers will become ineffective.

Momentum Exchange Electrodynamic Reboost Tether (MXER Tether)

The story of David and Goliath tells of how a small boy uses a sling, a small rock, and a little momentum to defeat a giant problem. This is similar to a MXER tether. MXER is a combination of two tether concepts. The Electrodynamic portion is discussed earlier so lets move on to the Momentum Exchange component. David took a strap of animal hide, looped it, placed a rock in the loop, whirled it around to gain momentum, then released one side of the strap to free the rock. The rock took all of the momentum and flew out at a great speed to defeat the giant. In space a MXER tether would be a long cable with ballast at one end and netting at the other. The entire spacecraft might be 100 kilometers in length or more. The tether would orbit the Earth and spin like the spokes of a bike.

Once the tether is in place and travelling in an elliptical orbit, a satellite would be launched into LEO. The tether would swing a catch mechanism down and snag the satellite. Just as David threw the stone, the tether would rotate and release the satellite to propel it to a new and higher orbit. The tether would, in effect, transfer

An elevator could be used to modify the tether ballast.

Solar panels provide power for tether systems and the electricity needed for the metal wire in the tether itself.

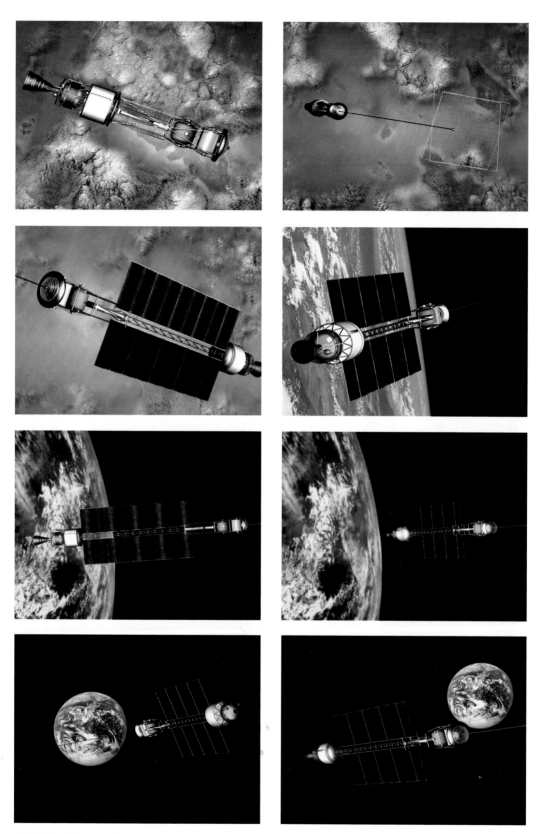

MXER tether deployment sequence.

the momentum it had gained to the satellite.

When the tether transfers some of its momentum to the satellite, its own orbit is lowered. Without some reboosting it would eventually deorbit. To raise the tether's orbit, electrodynamic principles are used. The long tether will have a metal wire running through it. Driving an electrical current will create the gentle push needed to reboost itself. This push is soft and may take several months of reboosting to reach the desired orbital momentum needed for the next momentum exchange to a satellite.

Typical Components of a MXER Tether

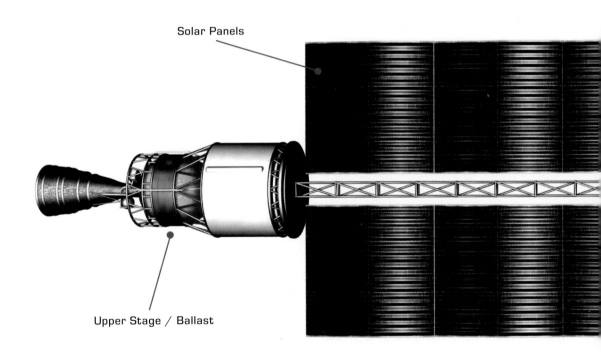

Solar Panels

Upper Stage / Ballast

The fantastic thing about tethers is that they work on pure physics. The MXER tether does not require refueling. Tethers have been flight-tested in the past and some technical challenges still exist. The electrical current of the tether has been significant enough to short out and snap the tether fibers. A 100 kilometer tether would stand a great risk of being hit by space debris or micrometeorites. New multi-filament tether cables are being developed that will allow for many cables to be cut at various locations and still allow the tether to retain its required strength.

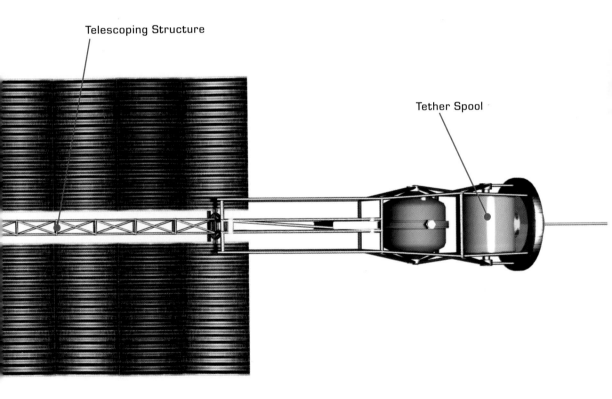

Telescoping Structure

Tether Spool

Aeroassist Maneuvers

Aeroassist is a term used to represent various methods to accelerate or decelerate a spacecraft using a planet's atmosphere. It uses the atmospheric gases to create drag or lift. The Space Shuttle uses aeroentry when it reenters the atmosphere. The process slows the descent and allows the vehicle to land. It creates tremendous heat on the spacecraft. Aerogravity-assist is used to swing a spacecraft around an orbiting body and to accelerate the spacecraft. To gain speed, a spacecraft will dip into the atmosphere and get an aerodynamic lift like an airplane to accelerate the vehicle. Aerobraking does the opposite of aerogravity-assist. A spacecraft enters into orbit and dips slightly into the atmosphere to slow down. The shape of the aerobraking spacecraft is designed to get maximum drag in the atmosphere where the aerogravity-assist design reduces it. Recently the Mars Global Surveyor used aerobraking to achieve the desired circular orbit needed for mapping the surface of Mars. Since aerobraking dips slightly into the atmosphere, the change is rather modest, and the whole process can take several months.

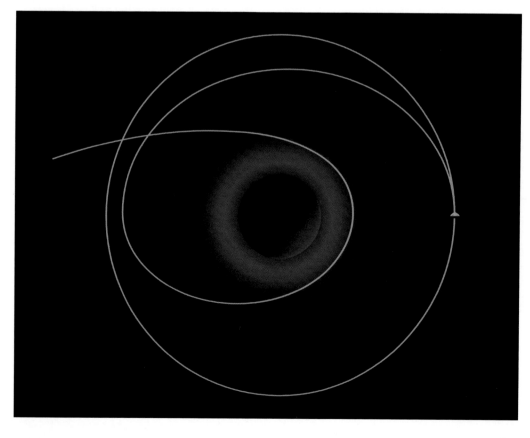

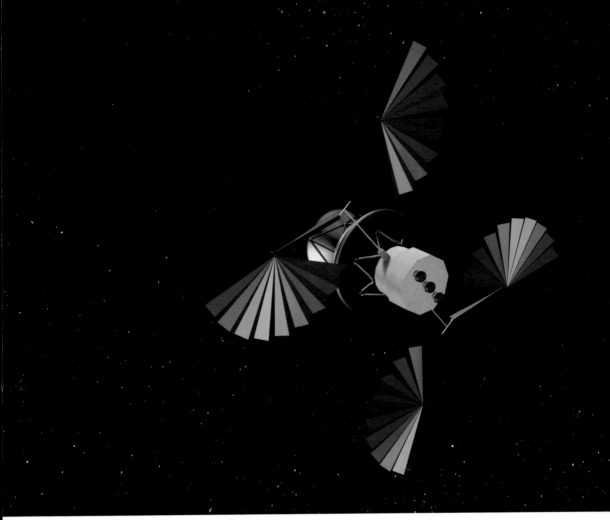

Solar array deployment.

Aerocapture

Time is a considerable factor in any mission so anything that can save time is worthy of study. A typical in-space mission strategy is to accelerate for one-third of the distance, coast for another third, then slow down at the last third. Just as a propulsion system is required to accelerate in the start of a mission, propulsion is generally needed to slow the space-craft for rendezvous with a planet. The spacecraft must slow down to enter into orbit around the destination planet or it will just fly by it very fast. This process of slowing the spacecraft can take years and take additional fuel. Aerocapture will use the atmosphere to slow the spacecraft into a capture orbit in a single pass. This will save travel time, as the spacecraft will not need to slow down as it reaches its destination. The spacecraft would rapidly slow down upon arrival.

The spacecraft will swing deeply into the atmosphere and get maximum drag. This will absorb much of the speed, dissipating it as heat, and put the spacecraft into orbit. With a couple of modest orbit correction maneuvers the spacecraft will be in

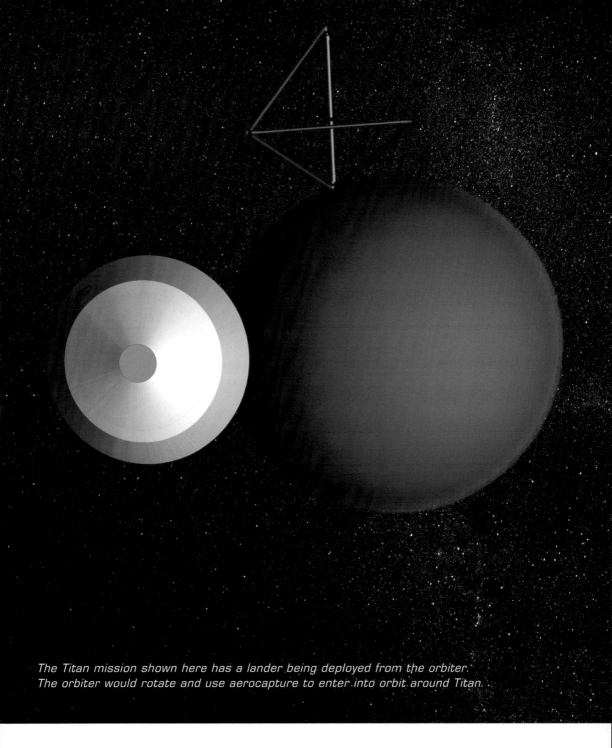

The Titan mission shown here has a lander being deployed from the orbiter.
The orbiter would rotate and use aerocapture to enter into orbit around Titan.

a circular orbit in just a few orbits. Aerocapture represents a significant time saver in overall mission time.

One of the biggest challenges facing an aerocapture spacecraft is knowledge of the destination atmosphere. Where aerobraking is dipping only slightly into the atmosphere, mission controllers can quickly make adjustments. Since an aerocapture craft goes much deeper, it has much less opportunity for correction. Furthermore, the speed of entry is much greater and further complicates any corrective actions. For these reasons it is critical that mission planners have as much information about the atmosphere of the target planet as possible. Much is known about Earth's atmosphere as well as that of Mars. One of Saturn's moons, Titan, on the other hand has not been explored. It is known that Titan has a dense nitrogen-rich atmosphere but engineers will need as much information as possible to properly plan and design a mission to Titan using aerocapture.

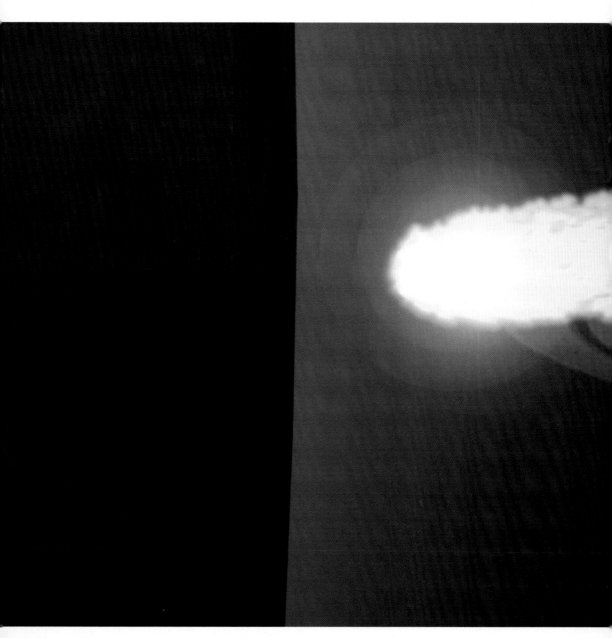

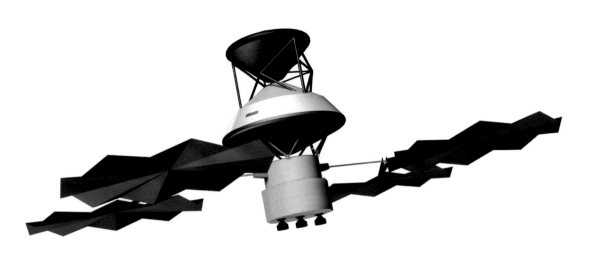

The spacecraft is designed to use the atmosphere to slow down. An enormous amount of heat energy is created. Effective protection from the heat build-up is critical for aerocapture's success.

Solar Electric Propulsion (SEP)

The Sun is commonly used as a source of power in space. Sunlight is a dependable power source in space because it is intense and plentiful as long as you are relatively close to the Sun. The International Space Station (ISS) has huge solar panels to power services, charge batteries, and power experiments. The solar arrays generate enough energy to power 200 households down on Earth. Batteries are used to power the spacecraft when it is in shadow. The space station is in shadow for about 20 minutes of each 90-minute orbit. Solar Electric Propulsion (SEP) uses the electricity generated by the solar arrays to power electric thrusters. Solar panels work great while there is sunlight to power them. In Earth orbit there is ample sunlight to provide power for almost all spacecraft applications. Closer to the Sun there is more power, but it falls off quite quickly as you move outward in the solar system. As mentioned before, the distance from Earth to the Sun is about 93,000,000 miles and this is called one Astronomical Unit (AU). At twice the distance from the Sun, or 2 AU, there is only 25%

Typical Components of a SEP Vehicle

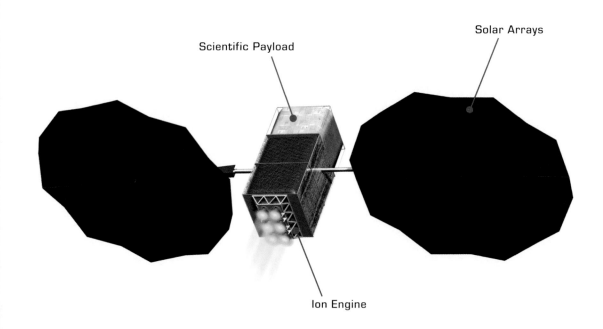

Scientific Payload

Solar Arrays

Ion Engine

Ion engines can operate for years.

as much light. For this reason, solar electric propulsion systems normally get all of their momentum (rocket thrust) while in the inner solar system and then coast to their destinations.

An example of an SEP application that is being considered is a follow-on mission to Titan after Cassini and Huygens. (Cassini, a nuclear powered spacecraft is destined to arrive at Saturn in 2004 and drop the Huygens probe onto Titan for a brief look.) In this new mission to Titan, the spacecraft would use an ion thruster propelled SEP system to get to Saturn and an aerocapture system would slow the spacecraft and put it into orbit around Titan. Getting to Titan is a tough job and the ion system could only be used out to about 2.5 AU. To help make the job a bit easier, the spacecraft would fly by Venus to get a "gravity assist."

There are other ways to generate electrical power in space besides the common solar cell panels. Concentrating sunlight with large mirrors can

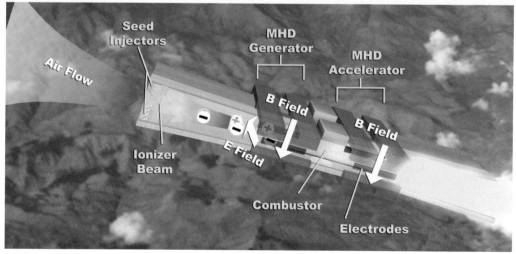

MHD engines accelerate performance by injecting electricity into the exhaust flow.

generate heat, which can be turned into electricity similar to how electricity is generated on the Earth. This system is called dynamic power conversion. The heat is absorbed into a gas and sent through a turbine which turns and drives a generator that produces electricity.

Several different types of thrusters use electricity. The most common is the ion thruster. This is an electrostatic device and is explained later. Two other types are the electrothermal and electromagnetic devices. Electrothermal thrusters use electricity to heat a propellant and send it out a rocket nozzle. The common names for these devices are arcjets and resistojets. They have been used in satellites for many years in space. The electromagnetic class of thrusters uses a combination of magnets and magnetic fields along with the electrical current. Some of the electric thrusters that belong to this class are MagnetoPlasmaDynamic or MHD, Pulse Inductive Thruster (PIT) and VaSIMR (Variable Specific Impulse Magnetoplasma Rocket). They are all advanced thrusters that are just now being studied and someday may be an important part of space transportation.

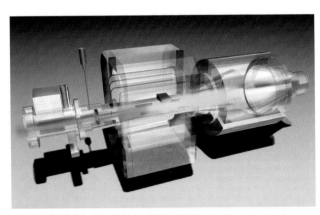

This test article would help scientists evaluate MHD accelerators using an arcjet.

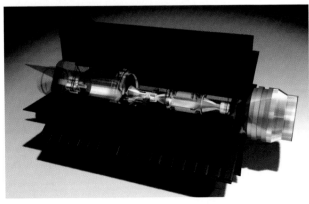

The VASMiR engine is a NEP that offers variable thrust.

Nuclear Electric Propulsion (NEP)

Nuclear Electric Propulsion (NEP) systems use nuclear power systems to generate electricity for thrusters like the ion thrusters discussed in detail next. In NEP-based propulsion systems, the nuclear reactor would generate energy that would heat up a gas which drives a generator. The generator would provide electrical power for the engines and other spacecraft operations. NEP systems have an advantage over SEP systems in that they don't rely on sunlight for power and thus can operate all the way to destinations in the outside solar system. Nuclear electric systems can power a spacecraft for years.

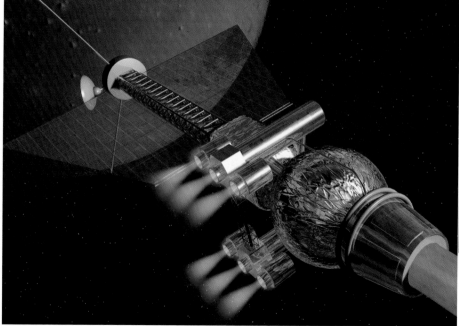

Radiators would remove excess heat from the reactor.

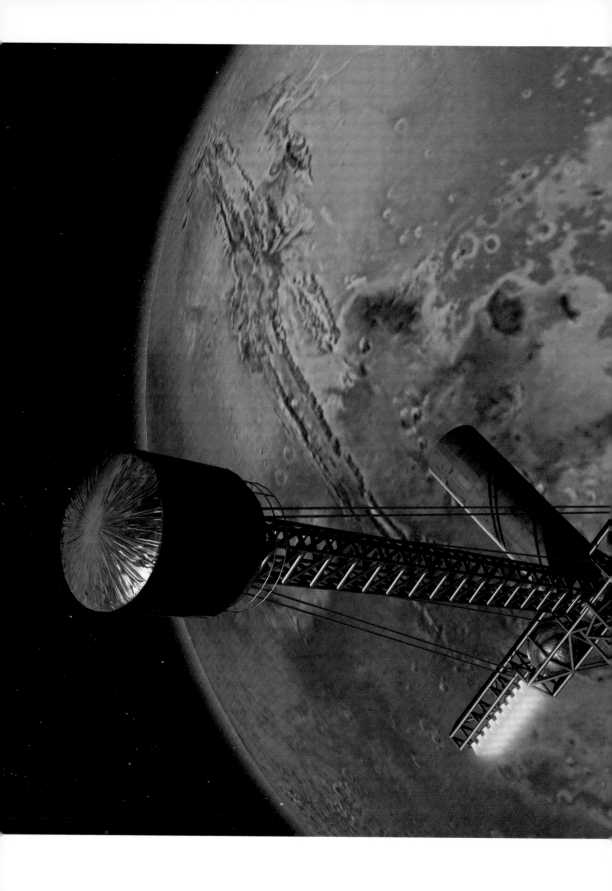

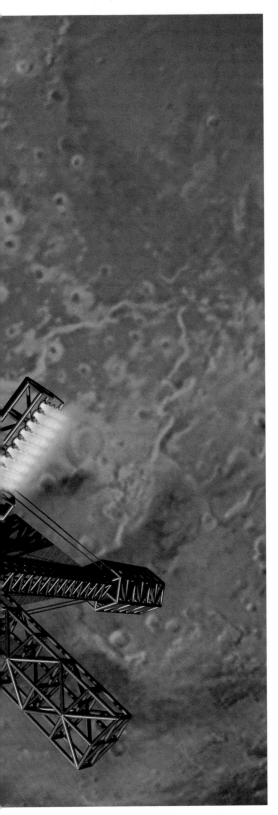

While NEP systems are more complex and expensive than SEP systems, they are required for very difficult missions like exploring the icy moons of Jupiter or going to the Oort Cloud in a reasonable amount of time. Nuclear systems also have the advantage of providing ample power for things like science experiments and communications once the spacecraft arrives at its destination. Nuclear powered spacecraft could also embark on scientific missions that would not be limited to a single destination. A nuclear powered probe might explore all of the moons of Saturn in a single mission for example. Nuclear power presents new potential for long-term exploration. The challenge for NEP will be in developing a dynamic power conversion engine that will endure a long time in the harsh environment of space.

The heat created by a nuclear reactor can be converted into electricity by turning a turbine. The two common power conversion methods are the Brayton and the Rankine thermodynamic cycles. Both take the heat and turn it into mechanical power that rotates an electrical generator.

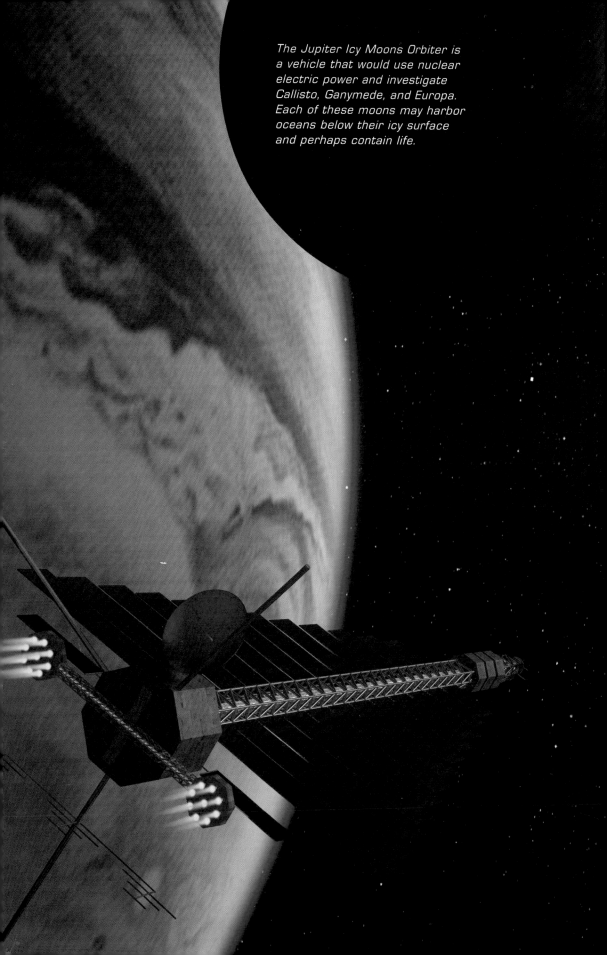

The Jupiter Icy Moons Orbiter is a vehicle that would use nuclear electric power and investigate Callisto, Ganymede, and Europa. Each of these moons may harbor oceans below their icy surface and perhaps contain life.

To do this they must reject the waste heat from the spacecraft. NEP systems all have big radiator panels that look a little like the solar panels on a SEP spacecraft. The radiator panels are pointed away from the Sun and out towards the cold dark background of space.

Normally, the nuclear power is derived from a fission reactor similar to the ones on Earth. Fission occurs when the uranium atom is broken in half by a neutron. When this happens, energy is released along with several other neutrons. Some of those new neutrons, in turn, hit other uranium atoms and split them, so the reaction continues building up heat energy. Radioactive isotopes are routinely used in space on most science missions in lieu of batteries. They do not use neutrons to fission uranium, but they spontaneously give off energy from their nucleus. Radioactive isotopes are usually very low power and fission reactors are high power devices. Fission reactors are easier to handle and safer since they are not radioactive when launched. Only after they are turned on and then operated for a long time do they become dangerous for humans to touch.

Fission reactors produce neutrons and gamma radiation that are often harmful to other parts of the spacecraft. To protect sensitive components or life on board the spacecraft a shield is needed in front of the nuclear reactor. The reactor is placed at the far end of a long boom with a thick shield next to it. The reactor, shield and power conversion system, which includes the large radiators, makes a heavy system. To be effective, the NEP engineers will have to design the system as lightweight as possible to keep the trip time to the outer planets under 10 years. Very large NEP systems (multi-megawatt) have been proposed for manned missions to Mars.

Ion Propulsion

Socks coming out of a clothes dryer sometimes stick together because they are electrostatically charged. As the socks bounce around in the dryer they rub together and electrons are knocked off and the socks become positively charged. This change at a molecular level causes a sock to be pulled toward another garment that is neutral or negatively charged. The invisible, molecular change can be stronger than the Earth's gravity and hold the sock to another garment. An ion engine uses this same basic force to propel a spacecraft.

An ion is created when an electron is knocked from an atom. The atom becomes positively charged as a result and is called an ion. In an ion engine, xenon or other gas is injected into a chamber where it is ionized. The ions are then accelerated out of the back of the engine. The force of the ion leaving the engine applies the same type of force as a conventional chemical combustion rocket engine. While a single ion will not push a spacecraft very far, an ion engine has streams of millions of them.

Note: This propulsion class is labeled an electrostatic thruster.

Gridded Ion or Hall thrusters are the two main types in this class and both have been successfully used in space. Since Gridded Ion thrusters are the most common, most people generally refer to them as "ion engines."

A fluorescent light or neon tube

is a vacuum tube filled with gas that excites molecules when electrical current is introduced. Electrons in the electric field strike electrons of the gas and knock them loose. The ions bounce around and emit light when they recapture an electron. An ion engine uses a gas like xenon because it has 54 electrons and it is relatively easy to ionize. The gas is injected into the engine's chamber, which is surrounded by a magnetic field. A cathode fires electrons at the xenon gas and ions are produced. A pair of electrically charged metal grids are at the opposite end of the thruster

Typical Components of a Ion Propulsion Engine

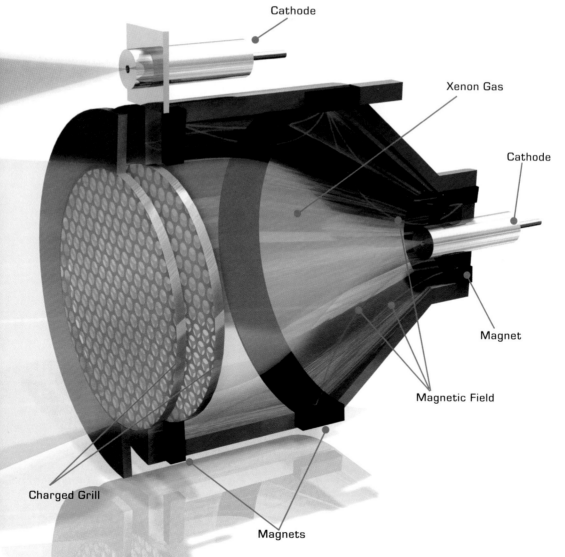

Cathode

Xenon Gas

Cathode

Magnet

Magnetic Field

Charged Grill

Magnets

chamber. One is oppositely charged and draws the ions by strong electrostatic force. The other grid is in front and has the same charge as the ion, only weaker than the outer grid. The first grid helps align the ion as it is propelled out of the engine at about 60,000 miles per hour! An additional cathode outside of the engine neutralizes the positive ions with electrons. Otherwise, the ion exhaust would build up outside the spacecraft and create problems.

At full throttle an ion engine can consume 2,500 watts of electricity and provide 1/50th of a pound of thrust. This is similar to the pressure a piece of notebook paper would have when it drops onto your hand. This is not much thrust and is hardly enough to push the vehicle quickly. The power of an ion engine is not in the amount of thrust it generates initially but that it can continue working for a long period of time at very high efficiency. Ion engines are designed to work for years. The result of this continual gentle pushing creates speed over months of continued operation. By contrast, chemical engines create significant thrust initially but only burn for a few minutes. Over time, the ion engine significantly out performs a chemical engine. An ion engine pushing for 20 months could accelerate a spacecraft like Deep Space 1 (DS1) to 10,000 miles per hour.

Ion engines were first tested on the ground in 1959. After 40 years of development an ion engine was demonstrated extensively (running for over 20,000 hours) in space on DS1. Initial tests demonstrated that the concept would produce thrust but it took time to develop an engine that would last. Development was slow because mission planners were not willing to take the risk associated with a new and unproven engine design. In 1992, NASA began the NASA Solar Electric Propulsion Technology Application Readiness (NSTAR) program. It was initiated to remove the barriers to using ion propulsion systems perfected for the New Millennium program. The New Millennium Deep Space 1 spacecraft launched with an NSTAR engine aboard a Delta rocket in 1998 and is still operational today.

Nuclear Pulse Detonation

A nuclear explosion has devastating power. It can level a city as history has shown. Much of the damage that a nuclear explosion causes is from a hypersonic shockwave that radiates from the center of the detonation. Nuclear pulse propulsion is a concept that dates back to the pioneering days of nuclear energy. Nuclear pulse propulsion attempts to utilize the energy of a nuclear explosion and translate it into space propulsion.

Project Orion, a classified military program in the late 1950's, was the first concept that embodied nuclear pulse propulsion. Project Gabriel was a small study to update the Orion design for use in a fast Mars mission or for asteroid defense. This is one of the featured vehicles on NASA's Starship 2040 exhibit. It not only uses fission energy as the Orion did, but fusion as well in its pulse units, since that is the only known way to practically obtain fusion energy in large quantities.

At the back of the Orion spacecraft is a blast plate attached to shock absorbers. The spacecraft has radiation shielding, a crew area, and a deployable lander. A series of small nuclear bomblets, or pulse, units would be ejected 100 to 1,000 feet behind

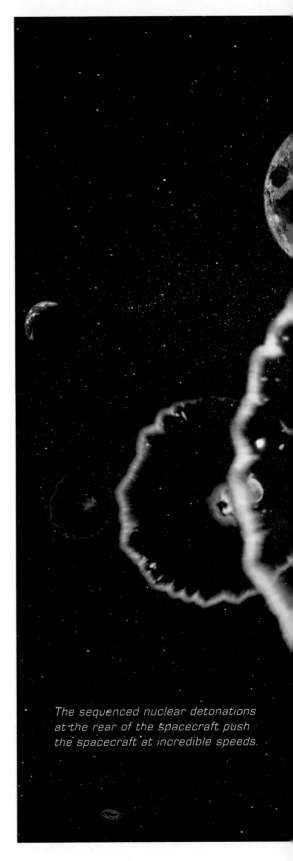

The sequenced nuclear detonations at the rear of the spacecraft push the spacecraft at incredible speeds.

the spacecraft and detonated. The explosion would impact the blast plate and push the spacecraft. Surfing on the shockwave, the detonations would occur at a rate of one every second to ten per second. This would propel Orion or Gabriel at incredible speeds.

Nuclear pulse could be useful for a variety of missions. Its speed and power make it an ideal candidate for manned missions to Mars. The speed would reduce the time and exposure for vehicle and crew to the dangers of space. The power would be useful for carrying the enormous payloads required for such a mission. Such a system would also be useful

Components of the Orion Spacecraft

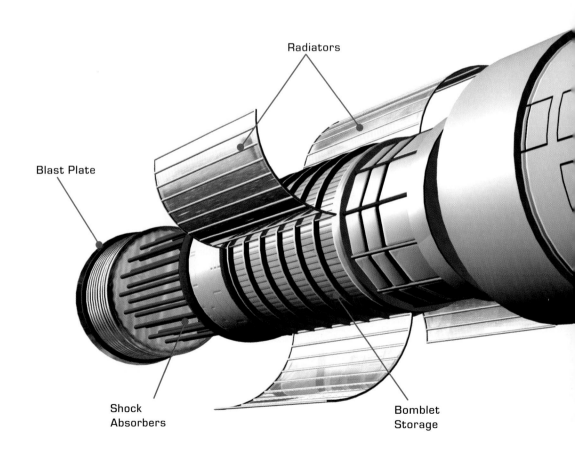

Radiators

Blast Plate

Shock Absorbers

Bomblet Storage

for deep space missions. Nuclear pulse propulsion would serve as a possible platform for manned missions beyond Mars or as a scientific probe that could carry and deploy scientific probes to a variety of destination planets or moons. Initial studies concluded that Orion could make a round-trip to Mars in 250 days and Gabriel in even less time.

The Orion program ended in 1965 after seven years of development and spending $11 million in research and development. Some scientists and enthusiasts still believe the Orion and concepts like it would be the most efficient use of nuclear energy for space propulsion.

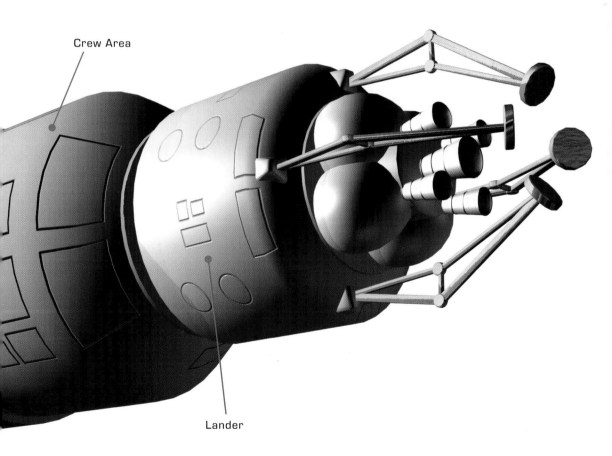

Crew Area

Lander

There are several obvious problems with this type of propulsion system. It is improbable that any nation would propose a spacecraft that carries and detonates nuclear pulse units for many reasons. Current treaties forbid nuclear explosions in space. Furthermore, if nuclear devices were detonated inside Earth's orbit it would interfere with or damage satellites already there. Therefore, the spacecraft must go beyond Earth-Moon space before 'going nuclear.' Then there is the issue of creating, testing and launching nuclear pulse units. It is unlikely that this would survive international pressures. Neither the United States nor any other nation would stand under the political objections or run the risk associated with the devices. The technology and the hardware, even though designed and built for peaceful purposes, would be desired by any group wishing harm to another. This would escalate the risk and expense and make development extremely difficult. The Orion is an intriguing, but very unlikely, spacecraft of the future.

Variations of this concept are still alive today. Some scientists propose this technology as the best possible asteroid shield. If an asteroid was detected and plotted to impact the Earth, the spacecraft would rendezvous with it in space as far out as practical. The spacecraft would fire pulse units to deflect its course to avoid hitting the Earth. It is much more realistic to try to push the asteroid off course rather than to destroy it outright. If the asteroid were to be exploded it is certain that some fragments, possibly large ones, would still impact Earth.

Nuclear Thermal Propulsion (NTP) And Solar Thermal Propulsion (STP)

Nuclear reactors generate a substantial amount of heat. Nuclear thermal engines use the reactor to directly heat a propellant gas and eject it out of a nozzle much like a chemical rocket engine. The nuclear reactor would heat hydrogen gas to around 3,000 degrees Kelvin. The molecules of pure hydrogen are much smaller than those of conventional chemical engine exhaust and exit the nozzle much faster. The result is about twice as much specific impulse or "efficiency" as chemical propulsion.

Nuclear thermal engines have been under development and consideration for more than 50 years. Both the United States and Russia have tested engine designs. The limiting factor has been the weight of the engines. The NERVA engine, tested over 30 years ago, was several meters in diameter and weighed several tons. In the 1980s a new design called a Particle Bed Reactor (PBR) was tested that would be lighter and smaller. The project held promise but was cancelled at the end of the Cold War. Development in nuclear thermal propulsion continues today but not at the past levels. The goal is to create a compact and lightweight NTP rocket engine and that has proven to be a great challenge because of material

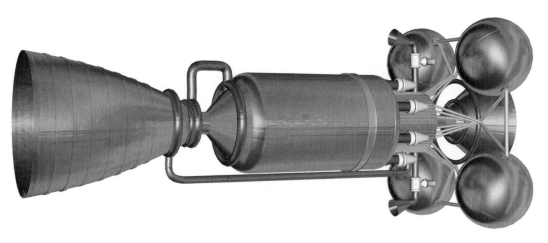

NERVA test engine.

temperature limitations.

Solar thermal propulsion (STP) is very similar to NTP in operation except the energy is from the Sun and is collected and concentrated by large curved mirrors. Temperature and material compatibility limits its performance as well. The inflatable lightweight mirror concentrators have been tested but not to the very large sizes required for a practical solar thermal rocket. Neither the NTP, nor the STP, system have been flown in space. Both use hydrogen as the propellant, which tends to make the propellant tanks large and unwieldy.

The solar sail would reflect sunlight to move through the solar system. The sail would be only 20 grams per square meter.

A Vision of Future Space Transportation

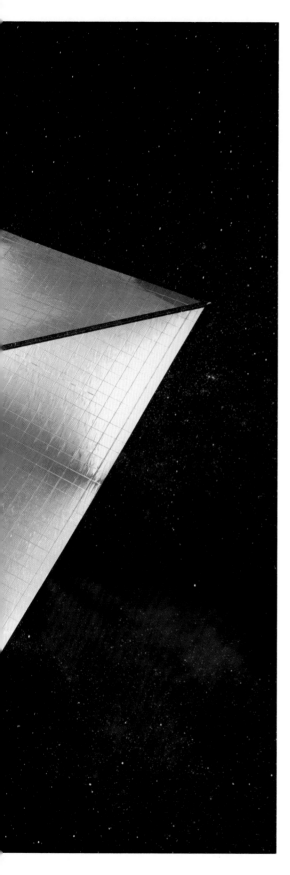

Solar Sail

The sail on a sailboat is pretty simple. Wind fills the sail and moves the boat. The hull of the boat keeps the vessel afloat and reduces the drag of the water. If the boat is on land where the drag is significantly greater it doesn't move an inch. With the right combination of sail size, ship weight, and hull design, the boat can sail. This combination creates a craft that moves using natural forces.

In space there is no air so how would a sail work? About 400 years ago Johannes Kepler noted the appearance of a solar breeze pushing and bending comet tails. More recently, at the turn of the twentieth century, American scientists Nichols and Hull, along with a Russian, Lebedev, successfully measured the gentle pressure that light exerts on an object. Building on this research in 1921, Ziolkovsky introduced the idea of a solar sail that would harness sunlight for propulsion. In effect, a sail rides on light beams.

Much like the sail on a sailboat, a solar sail has a large area to collect its energy. The pressure of the wind is much greater than the pressure of light so a solar sail must be much larger to catch as much light as possible. A typical square sail design would be a kilometer long on each edge. In Earth orbit, a reflective sail of this size would receive about 2 pounds of push. For a sailboat on the ocean this is equivalent to dead calm. A wind this slight would not have enough strength to fill a fabric sail, much less make it move. In space however, there is no drag so even the slightest push has some noticeable effect.

One of the advantages of having the Sun do the pushing is that it is relentless and free. No fuel is required. Sunlight gently pushes the sail again and again. Over time, the gentle pushes of the sunlight build up to create a significant thrust. The Sun is relatively intense near the Earth, but the farther you go in space the weaker the sunlight gets. Some sail missions envision a slingshot pass close to the Sun to build up tremendous speed for a trip into deep space. The sail would gain momentum while there is enough sunlight and coast when it is too weak.

While a small push can move an object in space, a solar sail must be incredibly lightweight to go somewhere fast. Current technology development activities are focused on developing high-strength, lightweight reflective sail material. Advanced Mylar materials are as light as 6 grams per square meter. Carbon fiber meshes have been developed to as light as 5 grams per square meter. Some predict that the fabric will need to be as light as 2 grams per square meter to be effective for deep space exploration. This is only the sail material; there must also be super-lightweight structural framework, guidance equipment and scientific instrumentation.

Weight is not the only challenge facing solar sail propulsion. A sail must survive the harsh environment of space. Space is full of intense ultraviolet radiation that will make most materials turn brittle over time. Much like the plastic rear window of a convertible top, traditional materials will crack and shatter. When exposed to the greater intensity of space, this effect is accelerated.

Space is a vacuum, but it's not empty. There are objects such as micrometeorites and other debris that can punch through a spacecraft.

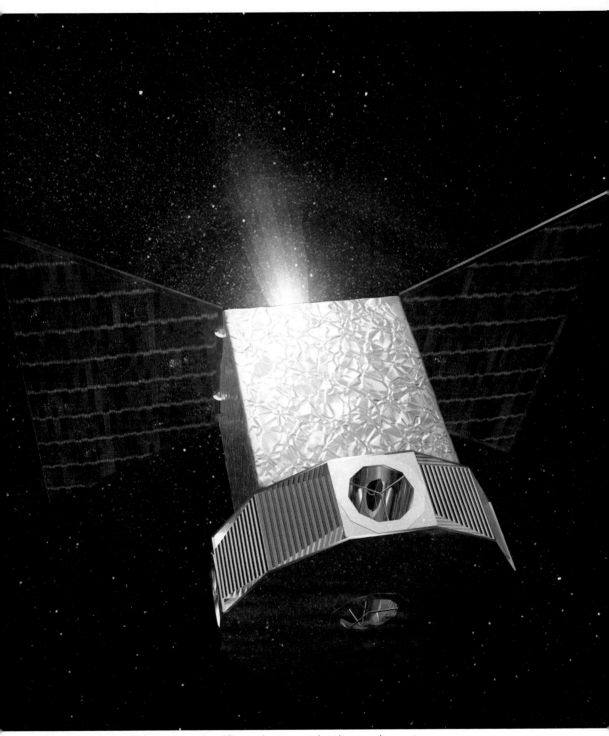

Solar panels would power scientific and communication equipment.

Since the surface area of a solar sail must be huge, it is that much more susceptible to being hit by another object in space. If the sail fabric is brittle or becomes brittle over time, it could shatter and end the mission prematurely. The sail surface must address the likelihood of hitting something in space and be able to continue the mission without interruption.

The payoff for solar sails is as great as the technical challenges facing them. The potential for scientific exploration is very exciting. A solar sail launched today would travel in eight years what it took Voyager 1 some 41 years to do. Scientists could explore the moons of Jupiter, the outer planets and beyond in a fraction of the time it takes today.

Solar sails aren't just for scientific exploration. Sails could be used as pole sitters. A pole sitter is a spacecraft that stays in orbit at either pole of the Earth. To remain in stationary orbit over the poles, the spacecraft would need a limitless propulsion system. A solar sail sitting at the pole could be used as a communications satellite or be useful for Earth observations and augment the existing fleet of satellites already in equatorial orbit.

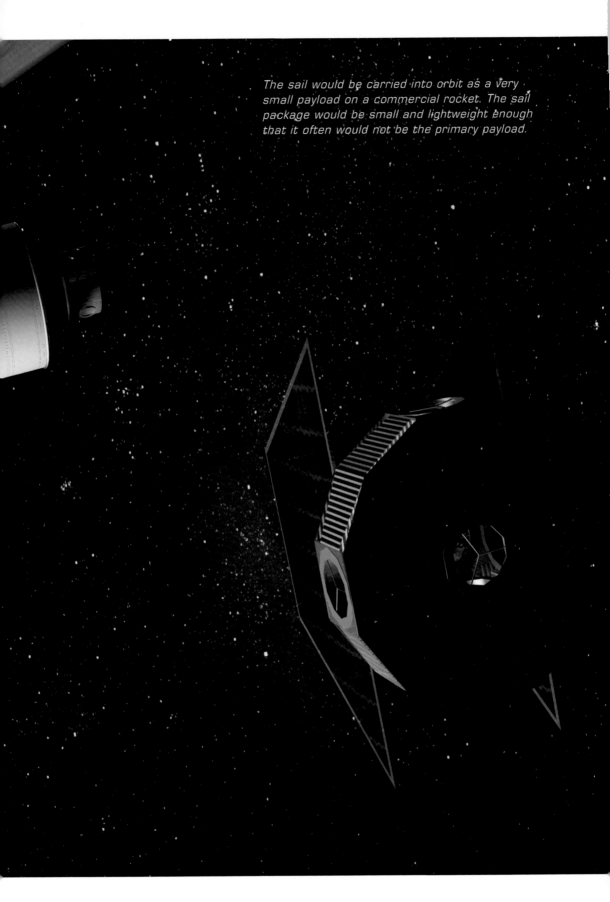

The sail would be carried into orbit as a very small payload on a commercial rocket. The sail package would be small and lightweight enough that it often would not be the primary payload.

Laser Sail

A laser sail works on the same basic principle as a solar sail. The laser sail is propelled by a laser beam rather than sunlight. The laser should provide higher acceleration than a solar sail. There are several components to a laser sail and they vary depending upon the concept. Each concept has a high power laser that directs a beam at a sail in space. If the laser were in space, it would point directly at the sail. A laser might also be on the Earth's surface. A laser on the ground would have a mirror in orbit that would reflect the beam to the sail. The sail would be a very lightweight disc shaped spacecraft. Initial studies predict that a laser sail could reach a speed of one-tenth the speed of light.

In 2000, experiments conducted at NASA's Jet Propulsion Laboratory (JPL) indicated that the basic concept behind a laser sail would work. Inside a vacuum chamber a laser was pointed at a piece of ultra lightweight carbon fiber. The laser was turned on and the carbon fiber moved. This represents the initial phases of research and much more has to be done before a laser sail system can be considered. Some estimates predict that as much as $80 billion dollars needs to be invested. This estimate can be compared to the cost of the Apollo Program that would be approximately $100 billion when compensated for current dollar values. This investment promises significant scientific payoffs. If the system launched 2 missions per year and functioned for 20 years, 40 star systems that are within 50 light years could be studied. Traveling at a tenth of the speed of light, a mission to Alpha Centauri would take about 43 years. This is only the time it would take for the sail to get there, it would take some significant time to transmit any images or scientific data back to Earth.

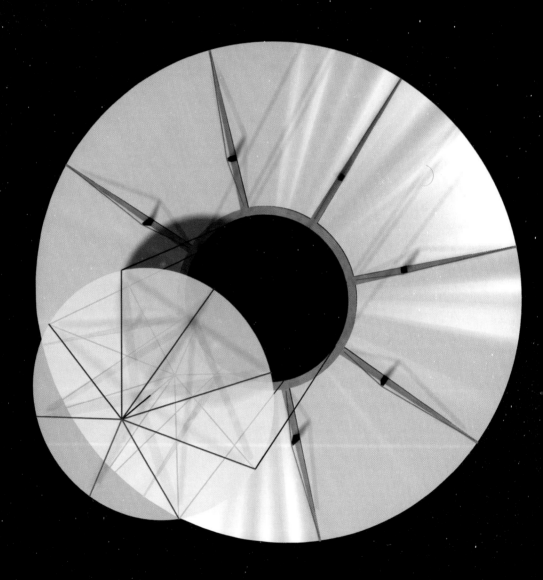

The center portion of the sail could
act as a brake to slow the sail and
be deployed around a distant star.

A very powerful laser would be required to power the system. For an Earth-based system a 26 giga-watt laser would be needed. To say this is a lot of electricity is a gross understatement. This represents about 12% of all of the electricity generated in the United States. This laser is for a lightweight scientific sail. Estimates for a sail capable of transporting a crew suggests that a 1500 tera-watt laser would be needed. Our entire planet only produces 13 tera-watts. A Saturn V generates about 0.04 tera-watts of power. It is needless to say that the laser power requirements represent one of the biggest challenges.

The lens would represent a significant challenge for both budget and technology. The lens would be made of thin plastic film. In order to be useful, the lens must be very large. Concepts envision a lens between 100 and 1,000 kilometers.

The sail would be made of ultra lightweight reflective materials such as carbon fibers that are currently being researched. The sail would be approximately 300 meters or larger in diameter. Some concepts envision a sail up to 1,000 kilometers. The sail might slowly rotate to help maintain its shape. Sail concepts envision a staged sail where an inner sail could separate and be deployed like a solar sail. As the sail reaches its destination star, the inner sail would separate from the outer ring. The laser would reflect off of the outer ring and be focused on the inner sail. This would slow the inner sail for rendezvous.

Laser sails are far term concepts. Technological advancements in light-weight sail material, nano structural materials, heavy launch systems, and giga-watt power generators must occur before laser sails can be deployed. However, laser sails promise fast interstellar exploration.

(The information in this section was derived in part from a paper entitled "Interstellar Flyby Mission Using Beryllium Laser Sail" by R.J. Halyard.)

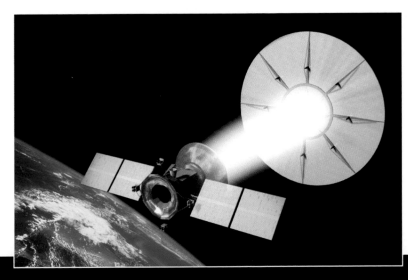

A laser would be located in space or on the ground and provide light energy to push the sail.

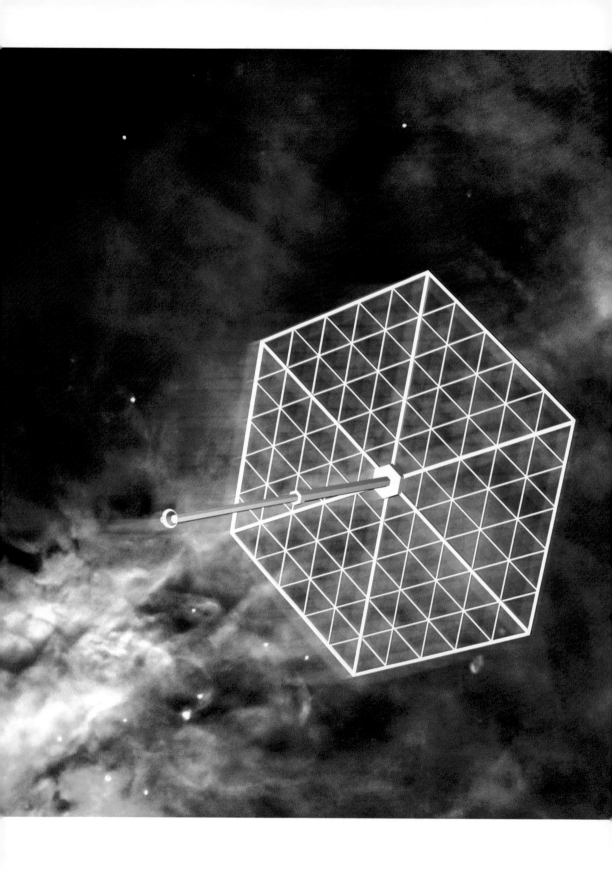

A Vision of Future Space Transportation

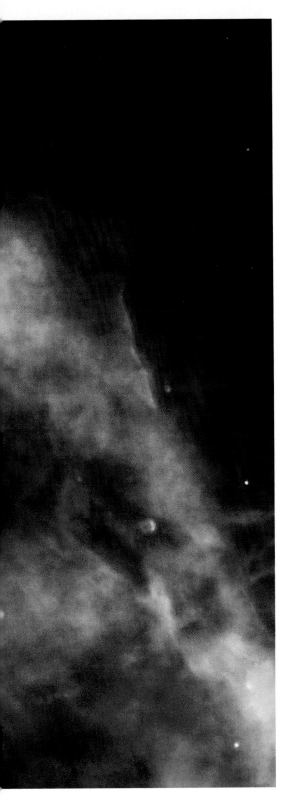

Microwave Sail — Starwisp

A microwave sail is similar to a laser sail. Microwave power is beamed from Earth to propel the sail. Unlike a solar or laser sail, a microwave sail is not a solid sheet but a mesh. The holes of the mesh are smaller than the wavelength of the microwave so the sail responds as if it were solid. The holes would help reduce the overall weight of the spacecraft.

The power requirements are significant. Approximately 65 giga-watts of power would be required. The sail would accelerate very quickly from Earth orbit. As the sail ventures deeper into space it would coast. At great distances, the microwaves would not be strong enough to push the spacecraft but would power the scientific instrumentations. Since the spacecraft is coasting, its missions would be limited to flybys. No rendezvous missions would be possible.

Plasma Sail — Mini-Magnetospheric Plasma Propulsion (M2P2)

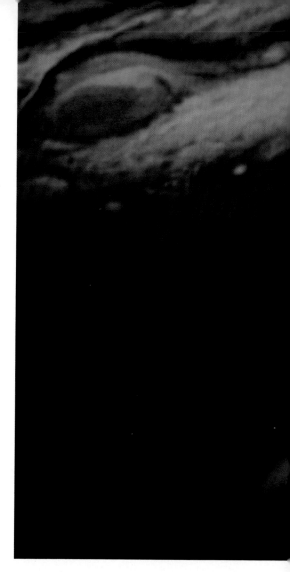

The Sun is the largest and most explosive feature in our solar system. The surface of the Sun has a temperature of 11,000 degrees Fahrenheit. Every second 700,000,000 tons of hydrogen is transformed into helium and 5,000,000 tons of energy is released. Like the Earth, the Sun has an atmosphere. The corona is the outer part of the Sun's atmosphere and routinely flares from the surface rip through and ejects massive amounts of energy and material through the solar system. These eruptions are called Coronal Mass Ejections or CMEs. On Earth, our atmosphere protects us from the effects of CMEs. The magnetosphere surrounding the Earth is pounded by the shockwave from the flares and most of the energy is deflected. Without this protection, life as we know it would not survive.

Energy from the solar flares move through the solar system at an incredible speed. This solar wind flows at velocities between 600,000 and 2,000,000 miles per hour. Solar wind impacts the Earth with great force but the effects are insignificant because of Earth's great mass. Imagine what might happen if the solar wind hit a much lighter object. This is the basic concept behind the M2P2 spacecraft.

The spacecraft would create a magnetic field then inflate it with plasma. This artificial atmosphere would be analogous to the one surrounding a planet. The artificial atmosphere might form a plasma bubble between 18 and 36 miles, yet the spacecraft would be very small and lightweight.

Solar wind retains its intensity

more than sunlight does since at twice the distance from the Sun, it still has half of its initial strength. As the solar wind hits the bubble it would deflect around it and push it along like a soap bubble in the wind. Some scientists predict that the M2P2 spacecraft could accelerate up to 180,000 miles per hour or 4,300,000 miles per day. In ten years, the spacecraft could travel 150 AU and reach the Heliopause.

Voyager 1, which was launched in 1977, will not reach it until 2019.

The M2P2 is a relatively new concept conceived by and under evaluation by Dr. Robert Winglee at the University of Washington. NASA has funded initial studies and further investigations are under consideration. The challenge is to demonstrate that the bubble will inflate as envisioned and the force of the solar wind can be controlled to propel the contents of the bubble.

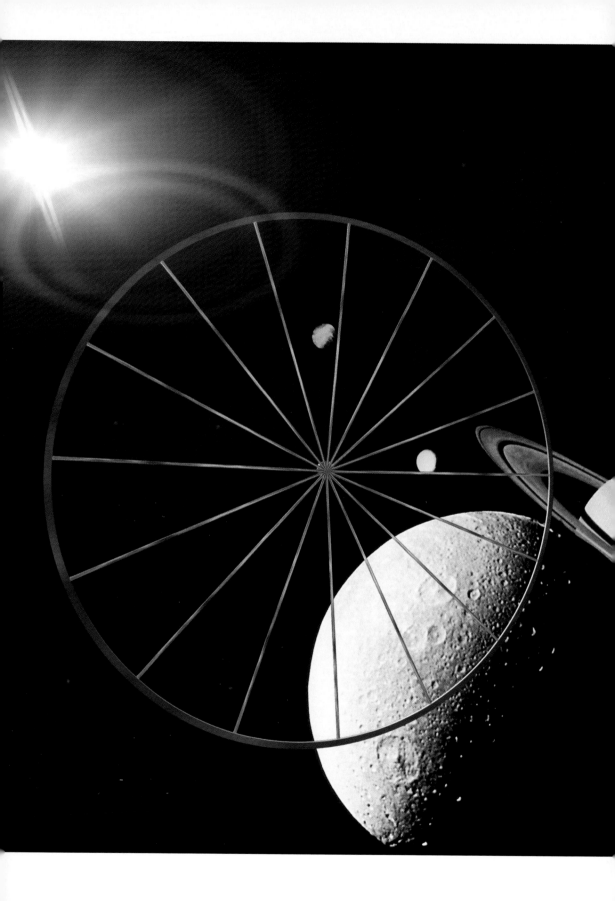

A Vision of Future Space Transportation

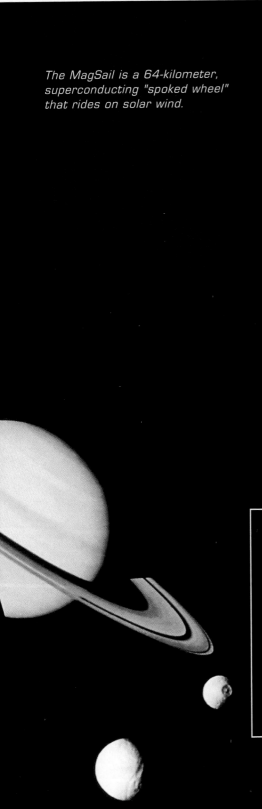

The MagSail is a 64-kilometer, superconducting "spoked wheel" that rides on solar wind.

Magnetic Sail — MagSail

Like the plasma sail, the magnetic sail will ride through space propelled by solar wind. As envisioned by Robert Zubrin and Dana Andrews, the MagSail is a 64-kilometer superconducting hoop with wires radiating from the center payload and vehicle systems module. The superconducting hoop creates a magnetic dipole that diverts the solar wind and thus pushes the spacecraft. The MagSail concept has a host of engineering challenges facing it. These challenges will require numerous engineering solutions and a multitude of technological advances. At the present time, the principle of the Magsail is well established, but its engineering implementation looks ominous.

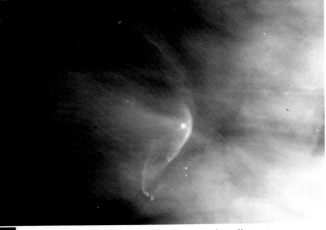

Bow Shock of solar wind around a distant star captured by Hubble.

*Enthusiasm finds the opportunities,
and energy makes the most of them.*

— *Henry S. Haskins*

Chapter VI
Private Initiatives

Chapter VI
Private Initiatives

The Space age began with a tremendous investment of public money. President Kennedy raised access to space to a national priority and unlocked the national treasury. After landing on the Moon, NASA built Skylab and a Space Shuttle. In 1984, President Reagan announced we would build and maintain a crew on a permanent space station within ten years. While the 1994 goal was not met, NASA still leads our nation's space effort. Guenter Wendt once told me that when he asked Dr. Wernher von Braun why he never took a job in the private sector for more money he was told that no one other than the government could afford to do what he wanted to do. (Mr. Wendt was on the launch pad for every lunar launch. His book *The Unbroken Chain* by Apogee Books tells his story.)

Dr. von Braun recognized that

bold steps in space carry significant risk and require substantial investments that often necessitate resources the size of the federal government.

While NASA is a government agency it actually builds very little space hardware. NASA typically contracts with engineering and aerospace firms to do the final engineering and cut the hardware. This influx of money from NASA and the Department of Defense has helped build the American aerospace industry. Government funds have fueled innovation and laid the foundation for future, private enterprises in space.

The thing that drives business is of course, money. In a free economy there must be a demand and a business model that promises a return on investment. Since space transportation is very expensive and risky, the development of new, advanced transportation systems will be very expensive and carry a high degree of risk. For investors in private industry the risks can outweigh the potential return. For private initiatives to be successful, or even get started, they must have significant funding and even more courage to face the potential of failure. Generally, the larger, main aerospace companies are too conservative to take this risk and tend to survive by working with the Government. However, there are several companies that are willing to explore the potential and take the risk. Some of the pioneering companies exploring the commercial future of space transportation include:

Bristol Spaceplanes

Bristol Spaceplanes is a British company that has three designs in mind. The Ascender is a test vehicle that would prove the concept for two other vehicles, Spacebus and Spacecab. Spacebus is a two stage passenger spacecraft. Spacecab is a scaled down version of Spacebus. Both vehicles would provide space trips to the public.

Pioneer Rocketplane

Pathfinder is a winged space plane concept by Pioneer Rocketplane that could deliver payloads to orbit. The Pathfinder spacecraft would have its LOX tanks filled in flight by a tanker aircraft. The RD-120 engines would fire to propel the craft up to Mach 15. Payload doors would open and launch the payload into orbit.

XCOR Aerospace

The Xerus vehicle is a small reusable craft that will provide brief trips into space for prospective space tourists. Xerus is a small winged body vehicle that will use rocket engines that burn non-toxic fuels.

Starchaser Industries

Starchaser Industries is a British company with an established history in the rocket industry. Two vehicles are envisioned that will service the space tourism industry. Thunderbird is a 52-foot tall vehicle and Nova is a scaled version of Thunderbird.

Scaled Composites

Proteus is a multipurpose vehicle. More an aircraft than a spacecraft, Proteus would serve as a first stage for launching small satellites. Proteus might also be a platform for space tourism and provide passengers with near-space experiences.

Space Access LLC

Space Access envisions a multi-stage air breathing spacecraft capable of delivering payloads to orbit. The first stage would look much like an airplane and launch the payload at high altitudes.

MBB Sanger II

The Sänger II is a German two-stage-to-orbit concept. The concept is named after the German scientist Eugen Sänger who proposed aero-space plane concepts before WWII. His wartime designs were not necessarily for peaceful purposes as you might imagine. The first stage would reach a speed of Mach 7 where staging would occur. The upper stage would launch and reach orbital velocity and would carry and deploy satellites or other payloads.

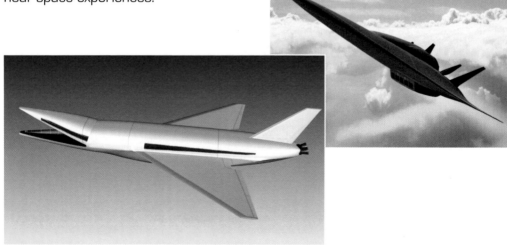

*The most effective way
to ensure the value of the future
is to confront the present
courageously and constructively.*

— *Rollo May*

Chapter VII
Current NASA Initiatives

Chapter VII
Current NASA Initiatives

It has been said that "the more things change the more they stay the same." Nowhere is this truer than at NASA. At NASA, change is a constant. With this in mind I realize that it is a bit silly to have a chapter entitled "Current NASA Initiatives." I realize that anything I write today stands a strong chance of being outdated in the near future. Furthermore, NASA has so many initiatives it is a considerable undertaking to summarize them all at any given point. With that said, I don't intend on summarizing NASA. Nor do I plan on describing a specific spacecraft "Program." Instead I would like to mention a current direction that NASA is working toward that I think represents very positive change with regard to the future of space transportation.

Next Generation Launch Technologies (NGLT) and In-space Transportation

Much of the imagery in this book was developed for NASA's Advanced Space Transportation Program (ASTP) at the Marshall Space Flight Center (MSFC) in Huntsville Alabama. This office looked to three primary areas: Earth to orbit, in-space propulsion, and fundamental research. ASTP funded research focused on technology for the next generations of launch technologies. The goal was to reduce the cost of access to space and enable scientific exploration of the solar system and beyond. This program has been recently redirected into NASA's larger Space Launch

Three stage to orbit design.

Initiative (SLI). The NGLT program now has two components, Earth-to-Orbit and In-Space Transportation.

The Earth to orbit component, NGLT, is tasked with developing the technology required to build and operate the next generation of launch vehicle. The near-term flight vehicle for NGLT is the development of a reusable first stage. This Reusable Launch Vehicle (RLV) will most likely be the first stage of a multi-stage mission. The RLV will glide back and land on a runway much like a conventional aircraft. Another primary focus of NGLT is the testing and evaluation of hypersonic air-breathing rocket systems. The X-43 activities represent some of the first flight-test activities.

This program represents the leading edge of research and development for our nations future spacecraft technology. NGLT is leveraging the best of NASA, the aerospace industry, and the nation's colleges and universities.

The In-Space Transportation component is dedicated to enabling further scientific exploration. In developing better spacecraft that will go faster and have more power on board, scientists will be able to explore more and learn more about our cosmic neighborhood. Faster spacecraft may enable us to explore the outer planets and possibly reach neighboring stars. Much of the in-space portion of this book is under their umbrella of interest.

Orbital Space Plane (OSP)

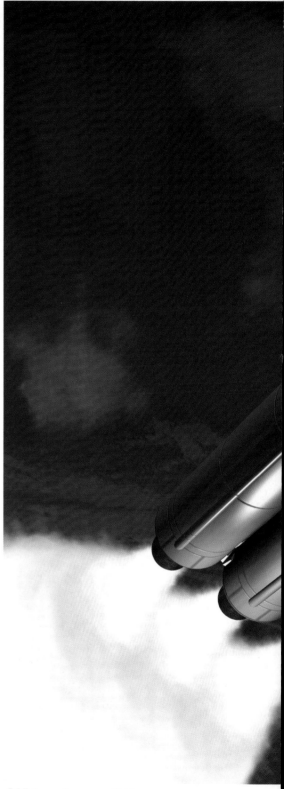

The International Space Station uses the Russian Soyuz and the Space Shuttle as crew return vehicles (CRV). The shuttle is certainly the first choice but it is not always docked or ready for launch. Space is a dangerous place and emergencies can occur anytime. Fire on the Russian Mir space station put the crew at risk. While risks of disaster are reduced at every opportunity, it is the real possibility of catastrophe that must be anticipated. The Soyuz will be used for years to come but a new CRV will be needed to ensure the safety of the ISS crew. NASA is working to develop an Orbital Space Plane that will meet this need.

The design for the Orbital Space Plane is not final at this time. It is anticipated that the OSP would be launched atop an existing expendable rocket such as the Atlas V or the Delta IV. The rocket would be spent during launch and not reused. The crew vehicle design might be a capsule, a lifting body, a winged body, or a sharp body design. The following illustrations represent some of the initial concept directions.

OSP launch on an ELV.

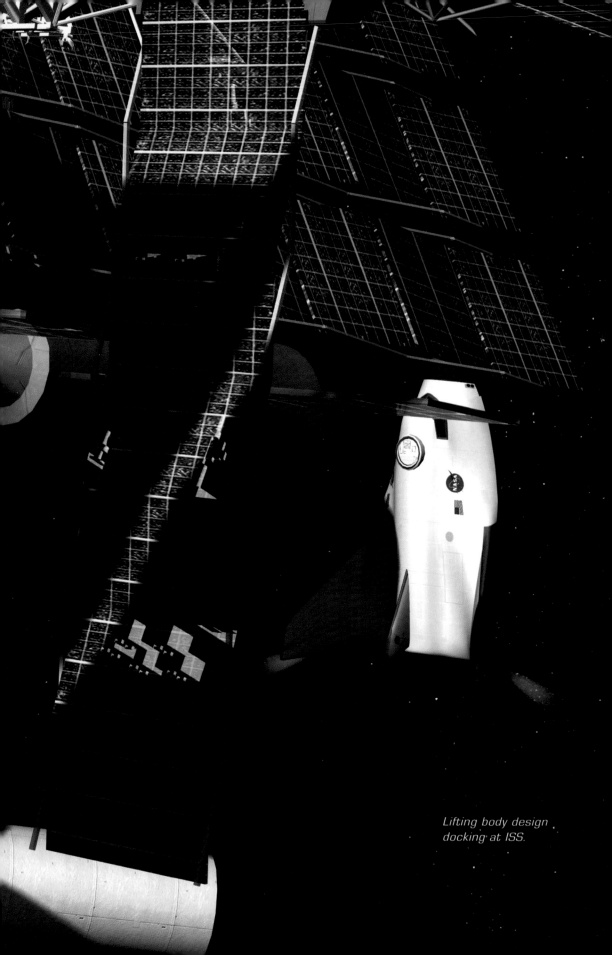

Lifting body design docking at ISS.

Capsule design.

Winged body design landing.

Lifting body at launch.

Capsule landing.

A Vision of Future Space Transportation

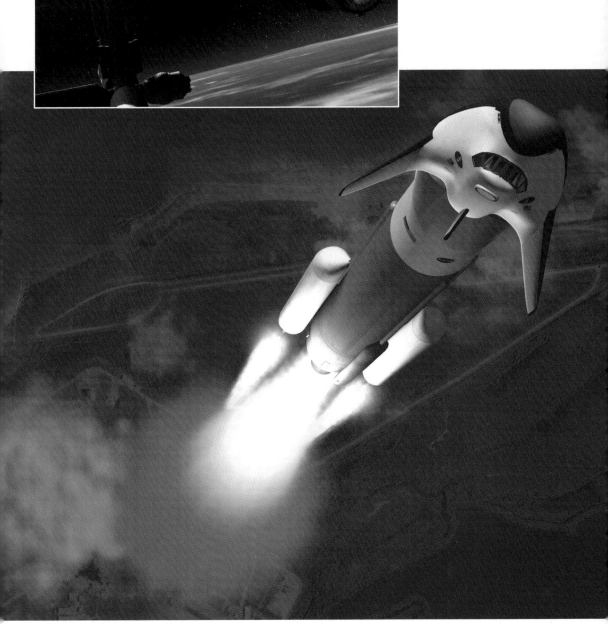

Capsule docking at ISS.

Lifting body on Atlas V ELV.

X-34 in flight.

Testbed Vehicles

Before NASA built the Saturn V and landed on the Moon, it had built and flown the smaller Redstone and Saturn 1B rockets among others. Before Neil Armstrong took the "giant leap for mankind," NASA took a series of careful developmental steps. Likewise, today NASA and the Department of Defense have been developing test vehicles to further the technology one step at a time. The lifespan of a testbed vehicle can be very short or quite drawn out. Some concepts never leave the designer's initial paper proposal, others have limited study, some have hardware manufactured, and some actually fly prototypes. While it is beyond the scope of this book to mention every test vehicle that NASA has in its portfolio, I will provide an overview of a few. The X-43 is one of the advanced propulsion vehicles and is presented in the ETO chapter. The X-37 is a different type of test vehicle that will help prove a unique set of technologies. The X-34 project was cancelled but was designed to test a new chemical engine as well as other subsystems.

X-34

The X-34 is a reusable testbed demonstrator that would be a platform for developing a suite of key technologies vital for the future of space transportation. The X-34 is an unmanned Reusable Launch Vehicle (RLV) that would be dropped from an L-1011 airplane, ignite its Fastrac engine and reach Mach 8. The X-34 would then return and land on a runway. The X-34 was to be a platform to test new Thermal Protection Systems (TPS), avionics, Guidance Navigation & Control (GNC), and other vital functions.

The X-34 was cancelled in 2001 before any powered flights were flown. A change in direction at NASA required additional safety systems to be integrated into the system after much of the research and development had already been conducted. This change required more expenditure than NASA was willing to fund, so the program was ended.

X-37

The X-37 flight demonstrator is envisioned to be carried to space in the cargo bay of the Space Shuttle or more likely atop an Expendable Launch Vehicle (ELV). Once in orbit, the X-37 will be controlled remotely from the ground. The X-37 will maneuver in close proximity to other orbiting spacecraft and demonstrate a variety of capabilities on orbit and in reentry. The X-37 will provide the advanced engineering foundation and the next technological steps needed for projects such as the Orbital Space Plane.

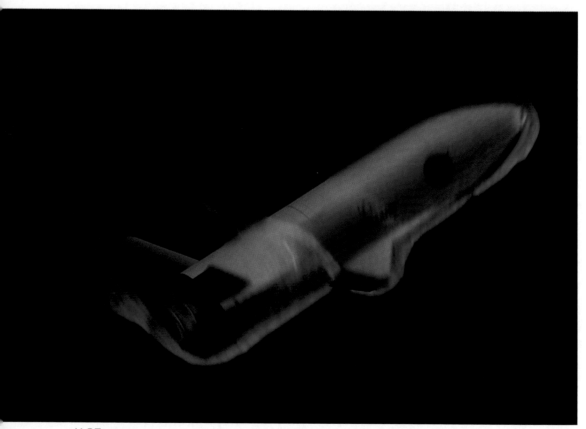

X-37 reentry.

The X-37 vehicle might be carried into orbit by the Shuttle.

Remote operations on orbit.

A Vision of Future Space Transportation

Autonomous landing.

X-37 deployment.

Children are likely to live up to
what you believe of them.

— Lady Bird Johnson

Chapter VIII
Space Day Foundation

Chapter VII

Space Day Foundation
Helping Kids To Reach For The Stars

Exploring the unknown is one of the most compelling human instincts. It is in this spirit that our space pioneers landed on the moon, peered at the farthest reaches of the universe, and introduced technologies that have transformed our lives. It is their remarkable vision that inspired the creation of the Space Day^{SM} Foundation, Space Day^{SM} and the Space Day^{SM} *Design Challenges*.

The Space Day Foundation

At the start of the new millennium, there are scientific, technological and economic challenges that call for a highly skilled, technologically literate workforce. There is a need to prepare students to meet the challenges of the new millennium. One factor that is often missing from our educational equation is the kind of real-life, hands-on involvement that can stir students' imaginations and inspire them to participate in the scientific process. The Space Day Foundation, in collaboration with Challenger Center for Space Science Education, sponsors the Space Day *Design Challenges*. The *Design Challenges*, an innovative educational program linked with the wonders of space and space flight, provides a much needed learning opportunity.

The Space Day Foundation, a 501(c)3 operating foundation, supports education initiatives, which use space as a motivating vehicle, aimed at contributing to the goal of inspiring student involvement in science, technology, engineering and mathematics. The Foundation's education initiatives are aimed at preparing students for the excitement and challenges faced by the workforce of the future. Student achievement is recognized and celebrated each year on Space Day. The Space Day Foundation serves as a forum for bringing together the best and brightest minds from business and industry, government and education to focus resources and efforts to achieve these strategic goals.

The Space Day Foundation is guided by a distinguished Council of Advisors composed of leaders in the fields of aerospace, education, science, business and government. Co-chairing the Council are Senator John Glenn and Dr. Vance D. Coffman, chairman and CEO of Lockheed Martin Corporation — both of whom consider the advancement of science, technology, engineering and mathematics education to be a national priority. Senator Glenn and Dr. Coffman also serve as co-chairs of Space Day, the award-winning educational initiative created to celebrate the achievements of our space pioneers and inspire the next generation of inventors and explorers to follow their lead. Among this annual event's flagship programs is the Space Day *Design Challenges*, which provides an exciting learning experience for students as they team up to solve real-life problems facing space explorers today.

Space Day[SM]

First launched in 1997, Space Day — which takes place each year on the first Thursday of May — is a massive grassroots effort dedicated to the extraordinary achievements, benefits and opportunities in the exploration and use of space. International in scope, this initiative involves thousands of teachers and students in dozens of countries around the globe.

Highlighting every Space Day celebration are local activities in schools and communities. Simulated space missions; special appearances by astronauts, educators and space experts; planetarium shows and museum exhibits; field trips and space games all reflect the enthusiasm and energy this annual event inspires. Space Day is also a forum for honoring the year-long achievement of students. Mayors and governors across the U.S. and Canada have issued official proclamations recognizing the importance of this unique initiative.

Sharing in the adventure is an ever-expanding array of active, enthusiastic Space Day Partners and Associates. These diverse groups, including local school districts and policy-makers at every level of government, members of the business community and international education, youth, space and science organizations, play a crucial role in bringing the excitement of Space Day to their constituents, members, colleagues and peers.

Space Day is designed to motivate students to explore their universe, offering them the chance to solve problems creatively. At the center of this effort is a series of Space Day *Design Challenges* developed by Challenger Center for Space Science Education. Interdisciplinary in nature, this exciting program encourages children to think like scientists and engineers. They are engaged in solving modern space-related problems, employing concepts in science, technology, engineering and mathematics as part of their design project. Teachers and students use the Web to connect directly with some of the best minds in science and engineering from government, business and education. For information on how to participate in the *Design Challenges*, visit the Space Day Web site at: www.spaceday.org.

According to Senator Glenn, "So much of what we have accomplished in space has been a potent combination of curiosity, imagination and the human spirit ... Space Day gives young people an opportunity to celebrate these pioneering achievements and embrace a future of endless, exciting possibilities." Brian Jackson, a teacher and regional coordinator for this initiative in Canada, concurs, "There is no better way to help kids discover the wonders of space exploration than through the activities sponsored worldwide during Space Day. Our future in the stars is more secure because of the interest that kids leave with on Space Day."

Space Day[SM] Design Challenges

Launched in 2000, the *Design Challenges* program is a set of hands-on projects that allow students to investigate, research, design, and build unique solutions to challenges similar to those encountered in the space program.

The *Design Challenges* were developed by Challenger Center for Space Science Education, an international not-for-profit organization created in 1986 by the families of the astronauts tragically lost during the Challenger space shuttle mission. In keeping with the educational spirit of that historic mission, the *Design Challenges* hope to inspire upper elementary and middle school students to pursue the study of science, technology, engineering and mathematics, while helping them develop teamwork, communication, and other critical life skills.

Thousands of students around the world have participated in the *Design Challenges* and benefited from this highly effective learning project. Each year, students are asked to develop their own solutions to a new set of challenges, such as:

X-treme Fitness — To help the astronauts living on the space station retain bone mass and muscle tone, create an entertaining sport, game, or piece of exercise equipment for them.

Space Walk Talk — To develop a nonverbal way for the astronauts on extravehicular activities (EVAs) to communicate with the crew inside the International Space Station.

Inventors Wanted — To help astronauts exploring Mars invent, design, and build an item that could make living or working on Mars easier and/or more enjoyable.

The Process

"The Challenges are carefully designed to emphasize process. The end products were exciting to the students — the way they handled themselves in collaborative groups was exciting to me as a teacher."
— Karen Godenschwager, University School, Shaker Heights, Ohio

Students begin by asking questions, such as: What design features does an airplane need to fly on Mars or another planet? What is the space environment like? How are the environments of Earth and Mars similar? How are they different? How do humans live in space? What is the difference between living in space and on Earth? How can we build tools to explore planets in our solar system? Questions like these allow students to focus their research. Once their research is complete, they begin conceptualizing their solutions. They have drawn diagrams, gathered materials, designed experiments, built models, etc. Once satisfied with their solutions, they are encouraged to test and evaluate them to determine whether their solutions work as anticipated.

The *Design Challenge* makes the scientific process fun and encourages active participation. By showing students how the "real world" works, they can see for themselves that jobs involving science and mathematics can be rewarding.

Standards-based Content

"Students were participating in 'real life' problem-solving situations which helped make the learning more meaningful, and kept motivation high." — Catherine Creech, C.R., Applegate Elementary School, Freehold, New Jersey.

Science, technology, engineering and mathematics, are critical subject areas in education today. Space is an exciting topic for students. Combined, education and space form a powerful means to engage the imagination and excite students about learning. The *Design Challenges* builds from the national education standards in these subjects, as well as in language arts, fine arts, and social studies to give students a fun, yet meaningful project.

Using the scientific process throughout the *Design Challenges*, students must first determine what they need to learn in order to develop their solution. For example, students first researched the differences and similarities between Earth and other planets in our solar system before designing robotic missions to these planets. Others studied human physiology and microgravity, and designed fitness equipment to keep astronauts healthy and strong. In 2003, students explored the principles of flight, such as lift, drag, and thrust. They researched how a jet engine works. They investigated what aerospace engineers are working on now that may change flight as we know it.

Teamwork and Collaboration

"The best part...is that [the Design Challenges] allowed my students to work together cooperatively on a challenging project. It required students to accept certain positions on a team and made them responsible for a particular job."
— middle school teacher from Paradis, Louisiana.

To be successful in almost any field, you have to be an effective member of a team. The *Design Challenges* stresses teamwork and collaboration in order to help students build this critical life skill. As students work on a solution, they form a team and take on a Leadership Role, such as Design Engineer, Team Ambassador, or Lead Scientist. Each member has specific responsibilities for their role, in addition to conducting research, providing ideas for the design, and helping to build the team's solution.

Students collaborate with other teams by asking questions, answering questions, providing encouragement, and sharing information. For students with Internet access, collaboration is encouraged through the monitored discussion boards provided by ePALS Classroom ExchangeTM. These boards let students "talk" to students around the world. Collaboration can also occur between teams in the same class or school.

Collaboration goes beyond student-to-student interaction and includes discussions between students and professionals in the field. Through the discussion boards, students can interact directly with experts to ask questions and receive encouragement as they work on their projects. Featured guests have had expertise in such diverse areas as robotics, engineering, aerospace, and journalism.

Integrating Technology

"The challenges promote hands-on application of their [students'] knowledge in science and in technology. They had fun."
— Maureen Armbruster, Suffield Elementary School, Mogadore, Ohio

The *Design Challenges* allows for seamless integration of technology into the classroom. The program provides students with a variety of ways to use both computer technology and technological tools. For online resources, the *Design Challenges* offers moderated discussion boards, as well as Web casts, Web chats, and/or videoconferences offered by Space Day Partners, such as NASA's

Johnson Space Center.

Technology is available to students in many forms — the *Design Challenges* gives them creative ways to use it. They use technology in the design of their projects and to communicate the results of their work to others. For some *Design Challenges*, students submit digital photographs and videos of how their solutions work. Many enterprising teams have created their own Web sites that highlight their solutions, research, and team members.

National Recognition

Students who participate in the *Design Challenges* have fun learning new topics, working as part of a team and engaging in authentic science. Teachers are thrilled to see the creativity and ingenuity of their students. By participating in the *Design Challenges* and submitting their solutions to Challenger Center, students have the opportunity to be recognized for their hard work.

Design Challenges solutions are reviewed by the Space Day Education Advisory Committee composed of experts in the areas of science, engineering, education and journalism. Students submit their research, describe how they worked as a team, explain their solutions, and justify

their designs. The goal of the review process is to recognize the most outstanding *Design Challenge* solutions. Student teams have been recognized for "Best Collaboration," "Best Design," "Most Creative," "Best Presentation" and "Most Useful." Teams with the best solutions have been honored during the national Space Day celebration at the Smithsonian Institution's National Air and Space Museum in Washington, DC.

Space Day[SM] 2003...
Celebrating the Future of Flight

The theme for Space Day 2003 built on the celebration of the centennial of powered flight. Students were asked to design future flight systems under the thematic banner "Celebrating the Future of Flight." To celebrate the Wright Brothers' momentous achievement a century ago, the 2003 *Design Challenges* highlighted advancements in aviation and aerospace, and asked students to imagine where we will go in the future. The *Design Challenges* were:

Fly to the Future — Students envision, design, and build a model airplane of the future.

Planetary Explorers — Students design and build a model spacecraft that can fly on Earth and another planet or moon in our solar system.

Watt Power! — Students build a model aircraft that can remain airborne using a renewable energy source.

Teams also created illustrated short stories about their aircraft. They developed timelines of past events in flight and made predictions about future flight developments.

In designing their future flight systems, students were able to take advantage of moderated Discussion Boards that enabled students to pose questions to aeronautical engineering experts. Additional support was provided to the teams through the QUEST for EdVenture electronic lesson, a free satellite broadcast, focused on the process that has been used to develop aircraft from the Wright flyer to spacecraft. Students also participated in Web casts from Johnson Space Center and the World Space Congress, as well as a Web chat with aviation professionals portraying the Wright Brothers.

Participation in the 2003 *Design Challenges* was outstanding; teachers and students learned a lot about the scientific process through the development of aviation and flight designs. The Space Day Foundation and Challenger Center are pleased to recognize the teachers and students representing the "Stellar" *Design Challenge* Teams for 2003.

Chapter Contributors:

Ralph K. Coppola, Ed.D.
Selma Mead
Karen Offringa
Merri Oxley

Space Day Design Challenges
"Stellar" Team Designs for 2003

Teacher Name: Ronnie Voigt
Student Names: Kent Southern, Greg
                Pelletier, Connor Fogle, Cloe
                Ayenu, Abigail Marie Wells,
                Ashlee Smith, Corbin Farr,
                Samantha Weber,
School Name: Walkersville Elementary
City, State: Walkersville, Maryland
Design Solution: Planetary Explorers
Team Name: Techno Titans

Teacher Name: Karen Godenschwager
Student Names: Michael Eby, David
                Hrvatin, Kevin Krajewski
                Alex Kovalik, Charles Stone,
                Nick Smedira
School Name: University School
City, State: Shaker Heights, Ohio
Design Solution: Spacers
Team Name: Spacers

Teacher Name: Ms. Lissa Scarpellino
Student Names: Matthew R. Heon, Kelly
                Joyce Johnson, Sara
                Faciocani, Nicholas
                DeMarinis, Ethan Faciocani,
                Kathleen A. Bacon,
                Meaghan I. Bacon,
School Name: Rhodes Elementary School
City, State: Warwick, Rhode Island
Design Solution: BAT
Team Name: Fledermaus

Teacher Name: Joe and Lauren Cummins
Student Names: Jeremy Cummins, Zachary
                Cummins
School Name: Cummins Home School
City, State: Walton Hills, Ohio
Design Solution: Air Jams
Team Name: Team Bernoulli

Teacher Name: Jean Eyman, Kari Imlay
Student Names: Chumar Williams, Sam
                Imlay, Samantha L. Nguyen,
                Bhaskar Vaidya, Angela Jin,
                Maddy Levin,Claire Johnson
School Name: Franklin Magnet Middle
                School
City, State: Champaign, Illinois
Design Solution: CSSC-BAM V
Team Name: Team Jupiter

Teacher Name: Ms. Bethany Goerdel
Student Names: Kalyn Irvin,Kayla
                Casperson,Megan
                Trietsch,Jeremy C. Berg,
                Tyler Starr, Andrew Bray,
                Jake Mastin, Jared Ward
School Name: Justin Elementary School
City, State: Justin, Texas
Design Solution: JATOM-1
Team Name: JASA

Teacher Name: Mr. Edward Vawter
Student Names: Trevor Baier, Laura Webb,
                Timothy Webb, Evan Todd,
                Michael Vawter,
School Name: Grace Home School
City, State: Westerville, Ohio
Design Solution: X-76 Independence
Team Name: Young Ohio Engineers

Teacher Name: Ms. Susan Superson
Student Names: Dan Stearns, Jon Stack
School Name: Birchland Park Middle School
City, State: East Longmeadow,
                Massachusetts
Design Solution: M5-23 Fighter Jet
Team Name: Amber Team

Teacher Name: Mr. Ashley Kizer
Student Names: Caison Braswell, Samuel R.
                Love, Jack E. Feist, Jake
                Freeze,
School Name: Highlands School
City, State: Birmingham, Alabama
Design Solution: Stingray
Team Name: Night Hawks

Teacher Name: Ashley Kizer
Student Names: Jessica C. Simmons, Ella
                Agnew Sorscher, Anna
                Gray Sarcone, Danielle C.
                Chamoun, Lael Groover,
                Emma Harms, Aliisa
                Haikala
School Name: Highlands School
City, State: Birmingham, Alabama
Design Solution: Super Sneaky Space Craft
Team Name: Super Sneaky Girls

Teacher Name: Myron Landers
Student Names: Danny Gibbons, Rick Yablonski, Jonathan Colon, Jose Martinez
School Name: Electa Lee Middle School
City, State: Bradenton, Florida
Design Solution: Manta Ray Expedition to Mars
Team Name: Manta Ray

Teacher Name: Jan Smith
Student Names: Amber Wilmarth, Ashleigh Wilson
School Name: Art Freiler School
City, State: Tracy, California
Design Solution: Planetary Explorers
Team Name: A-dubs

Teacher Name: Cynthia Shoemaker
Student Names: Ryan Coomler, Cory Bryan
School Name: McComb Middle School
City, State: Hoytville, Ohio
Design Solution: Military Heavy-Lift
Team Name: The Ram Jets

Teacher Name: Elizabeth Wray
Student Names: Christian Kuylen, Zachary Betz,Jordan Ashcraft, Meghan Bilski, Mary Bernard, Jessica Marie Reid, Tyler A. Milano, Ryan James Manuel,Tori Bankston, Hannah Lynne Dees,Jude Patterson, Christopher Idland
School Name: Live Oak Upper Elementary School
City, State: Watson Louisiana
Design Solution: Arrow
Team Name: Lil' Rockets

Teacher Name: Teri Becker
Student Names: Jack Britton Pounds, Corey Neil Waldon, Kit Kunz, Eleanor O'Neil, Brian Reese, Alex Compton
School Name: Fredericksburg Elementary School
City, State: Fredericksburg, Texas
Design Solution: Aeromaniacs
Team Name: Aerohead

Teacher Name: Charlene Randall
Student Names: Victoria White, Carly Etter, Ryan Silva, Carina Michelle Hiscock, Harris Holley
School Name: Bowie/ALPS
City, State: Abilene, Texas
Design Solution: The Eagle
Team Name: NASA Kids

Teacher Name: Charlene Randall
Student Names: Christopher Campbell, Holden Gibson, Amber N. Reese, Lindsay Cranford, Marcus Andres, Gibson Aguirre,
School Name: Bowie/ALPS
City, State: Abilene, Texas
Design Solution: The Solarizer
Team Name: Firebolts

*I'm thoroughly convinced that
the first person to set foot on Mars
is participating in Space Day...*

— Dr. Sally Ride

It [the rocket] will free man
from his remaining chains —
the chains of gravity
which still tie him to this planet.
It will open to him the gates of heaven.

— Wernher von Braun

Appendix

Computer Imagery

A common question that we are asked is "How do you make the pictures on the computer?" When I first began working with computer animation on my 386/20 PC, I also found the process perplexing. Like many people, I could not see the path from the keyboard and mouse to a 3D graphic image. It is much like building a set for a movie and I will take you on a quick tour.

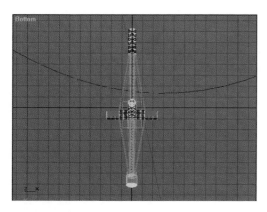

3D Modeling

The modeling software is much like a drafting table. The animator draws lines and shapes on the screen by clicking the mouse. As the animator draws, he is also working in 3D. Lines are drawn and projected into 3-dimensional space. Thus, the animator draws in virtual 3D space. Working like a sculptor, the animator constructs his object, rotates around it, and makes refinements. Everything that will be seen in the scene must

be created. If the animation is to rendezvous with the International Space Station then it must be created also.

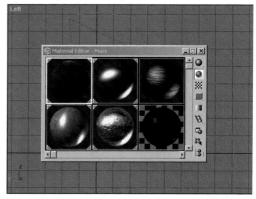

Textures

A 3D model is like a clay sculpture. It has no color and no texture. The animator must decide how the surfaces will look. If a surface is blue, for example, the animator sets the properties to the desired shade of blue. Scanned photos can be applied like wallpaper. Scanned textures might be decals, instrument panels, and starfields. Other properties include shininess, transparency, reflectivity, and so on.

Lighting

In the world around us light is taken for granted. We flip a switch and there is light. The Sun rises and there is light. In the animated world the animator has to set all of the lights. Lighting in the computer works much like the world around us. Light comes from a source, strikes an object, and casts shadows. The animator has control over virtually every aspect of the light: shadows, color, contrast, intensity.

Motion

With all of the models in place, textures applied, and lighting illuminating the scene, the scene is ready for movement. The animator decides the scene duration and positions the object over time. If a rocket flies past the camera the animator places it at the start position then rolls time forward and repositions the rocket. The software then moves the rocket between the two locations.

Rendering

So far the animator has been doing all of the setup but the image has not been created. With everything set up and tested, the computer processes all of the animator's decisions and paints the picture. The computer starts at the first frame and calculates how the light falls on the object's surface and how the surface properties should look. When one frame is completed, it moves on

to the next. The process can take minutes to hours per frame. For a second of video, the computer must create 30 animated frames. The process can take a considerable amount of time.

For the animator, the computer is a tool much like a chisel is to a sculptor or a paintbrush to a painter. While the computer software generates the image, or renders it, it is the animator who is the creator. The animator must observe the world around him and use the computer to create the image he sees in his mind. When the project is something that cannot be observed, the animator must use his creativity to create the imagery. Just as a word processor helps the writer, animation software helps the animator. Animation software doesn't make the artwork, it is just a tool for the animator. It is the animator who creates the artwork.

Acronyms

AAR - Air Augmented Rocket

ABLV - Air Breathing Launch Vehicle

ALTO - Spanish for high

ARC - Ames Research Center

ASTP - Advanced Space Transportation Program

AU - Astronomical Unit

CME - Coronal Mass Ejection

CRV - Crew Return Vehicle

CTV - Crew Transfer Vehicle

ED Tether / EDT - Electrodynamic Tether

EELV - Evolved Expendable Launch Vehicle

ELV - Expendable Launch Vehicle

ETO - Earth To Orbit

GEO - Geostationary Earth Orbit

GRC - Glenn Research Center

HTHL - Horizontal Takeoff Horizontal Landing

HTVL - Horizontal Takeoff Vertical Landing

ISS - International Space Station

ISTP - Integrated Space Transportation Plan

JPL - Jet Propulsion Laboratory

JSC - Johnson Space Center

KSC - Kennedy Space Center

LaRC - Langley Research Center

LEO - Low Earth Orbit

LeRC - Lewis Research Center

LOX - Liquid Oxygen

MEO - Mid Earth Orbit

MHD - Magnetohydrodynamic

MPD - Magnetoplasmadynamic

MLC - Microwave LightCraft

MSFC - Marshall Space Flight Center

MXER - Momentum Exchange Electrodynamic Reboost

NASA - National Aeronautics and Space Administration

NEP - Nuclear Electric Propulsion

NGLT - Next Generation Launch Technologies

NTP - Nuclear Thermal Propulsion

OSP - Orbital Space Plane

PIT - Pulse Inductive Thruster

RBCC - Rocket Based Combined Cycle

RLV - Reusable Launch Vehicle

RP - Rocket Propellant

SEP - Solar Electric Propulsion

SLI - Space Launch Initiative

SSC - Stennis Space Center

SSTO - Single Stage To Orbit

TBCC - Turbine Based
Combined Cycle

TSTO - Two Stage To Orbit

TPS - Thermal Protection System

VaSIMR - Variable Specific Impulse
Magnetoplasma Rocket

VTHL - Vertical Takeoff
Horizontal Landing

VTVL - Vertical Takeoff
Vertical Landing

Internet Resources

Advancespace.com

Aviationnow.com

Boeing.com

Apogeespacebooks.com

Howstuffworks.com

Interstellarsociety.org

Islandone.org

Jpl.nasa.gov

Jracademy.com

Msfc.nasa.gov

Nasa.gov

Slinews.com

Solarviews.com

Space.com

Spaceday.org

Spacetransportation.com

Spaceanimation.com

Spacefuture.com

Tethers.com

Tethersunlimited.com

SpaceAnimation.com is the online home for Media Fusion's space archives. From this site you can view imagery online and see some of the latest concepts. You can also order high resolution prints, DVDs, and other space related items. Check it out.

The Future of Space Animation: the CD-ROM

This interactive CD-ROM takes you beyond written text and images. You can watch animated videos created for NASA of concept vehicle missions. You can manipulate spacecraft concepts in full 3D view. You can even see selected Space Day[SM] Design Challenges "Stellar" Team designs fully illustrated.

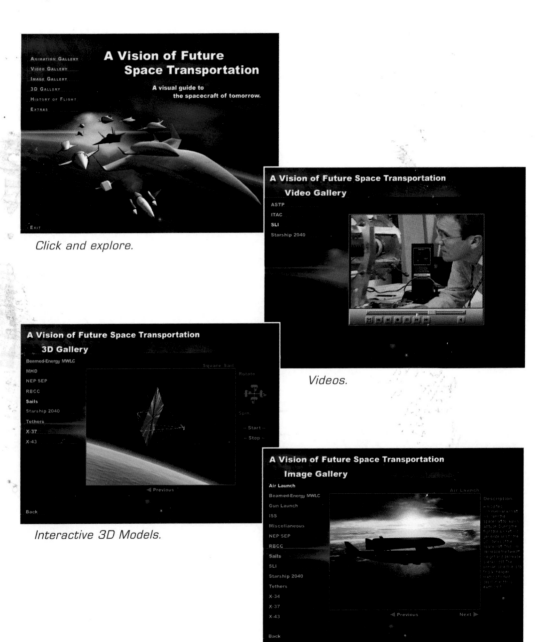

Click and explore.

Videos.

Interactive 3D Models.

Images.